Australia is an early adopter of new technology and Australians are usually receptive to practical innovations. Nuclear power is, however, the notable exception. Over the past half century, several inquiries have recognised the potential benefits and possible advantages of a local nuclear industry but a single nuclear power station has yet to proceed beyond the concept stage.

Submarines represent the most extensive application of nuclear power throughout the world, other than for industrial and household use. In 2016, the Australian Government announced that the 12 French-designed *Attack* class submarines replacing the ageing *Collins* class would be 'regionally superior' but conventionally powered. Nuclear propulsion was not considered.

This collection of thoughtful essays by highly experienced policy-makers, nuclear engineers, energy analysts and strategic planners considers the case for establishing an Australian nuclear industry, starting with the acquisition of nuclear-powered submarines to meet Australia's rapidly changing defence needs. The contributors call for an informed discussion of nuclear power that transcends the ideological rigidities of the 1980s and 1990s. Their insightful views provide a firm foundation for a continuing conversation the nation cannot avoid.

An Australian nuclear industry: starting with submarines? | ed. Tom Frame

ISBN: 9781922449382

Published in 2020 by Connor Court Publishing Pty Ltd

Connor Court Publishing Pty Ltd
PO Box 7257
Redland Bay QLD 4165
sales@connorcourt.com
www.connorcourtpublishing.com.au

Printed in Australia

Cover and page layout by Graham Lindsay

An Australian
NUCLEAR INDUSTRY
Starting with submarines?

TOM FRAME

editor

CONNOR COURT 2020

Contents

Contributors

Dr Ben Heard

Ben Heard is recognised as a leading voice for the use of nuclear technologies to address pressing global challenges. He currently works at Fraser-Nash Consultancy. After completing a Masters in Sustainability he started working in major projects in climate change, Ben was awarded his doctorate from the University of Adelaide in 2018, where he examined clean energy supply with a focus on nuclear technologies. He founded Bright New World in 2016 to provide a new organisation for people who want pragmatic, compassionate, and science-based environmentalism, in particular that values the role of nuclear technologies.

Dr Malcom Davis

Malcolm Davis joined the Australian Strategic Policy Institute (ASPI) as a Senior Analyst in Defence Strategy and Capability in January 2016. Prior to this he was a Post-Doctoral Research Fellow in China-Western Relations with the Faculty of Society and Design at Bond University from March 2012 to January 2016, and he currently retains an Honorary Assistant Professor position in the Faculty. He has worked with the Department of Defence, both in Navy Headquarters in the Strategy and Force Structure area, and with Strategic Policy Division in the Strategic Policy Guidance and Strategic External Relations and Education sections from 2007 to 2012. Prior to this appointment he was a Lecturer in Defence Studies with Kings College London at the Joint Services Command and Staff College in Shrivenham, 2000-2007.

Professor Lyndon Edwards

Lyndon Edwards is a senior executive at the Australian Nuclear Science and Technology Organisation (ANSTO) where he is the National Director, Australian Generation IV International Forum Research. He is a member of the Generation IV International Forum (GIF) Policy Group and GIF Expert

Group. For 10 years he was ANSTO's Head of the Institute of Materials Engineering. He is internationally recognised as an expert on Residual Stress and one of the leading proponents of engineering stress management using neutron diffraction. Professor Edwards studied for both his undergraduate and postgraduate degrees at Oxford University. Having previously been a Professor in the United Kingdom, he is currently Adjunct Professor of Engineering at Monash University.

Professor Tom Frame AM
Tom Frame joined the RAN College as a cadet midshipman in 1979 and served in the Navy for 15 years. He has been Anglican Bishop to the Australian Defence Force, a Visiting Fellow in the School of Astronomy and Astrophysics at ANU; Patron of the Armed Forces Federation of Australia; a Councillor of the Australian War Memorial and judged the inaugural Prime Minister's Prize for Australian History (2007). He is presently the Director of the Public Leadership Research Group and at UNSW Canberra and is the author or editor of over 50 books.

Dr Mark Ho
Dr Mark Ho works at ANSTO, Lucas Heights, specialising in reactor thermo-hydraulics. He is interested in reactor design, computational fluid dynamics, coding and boiling dynamics. He has attended meetings of the International Atomic Energy Agency (IAEA) in Vienna on Small Modular Reactors. He is the current President of the Australian Nuclear Association which is an organisation of professional scientists and engineers based in Sydney and with a branch in Adelaide.

Tony Irwin
Tony Irwin is a Chartered Engineer, Technical Director of SMR Nuclear Technology Pty Ltd and Chair of Engineers Australia Sydney Division Nuclear Engineering Panel. He worked for British Energy for more than 30 years, commissioning and operating eight nuclear power reactors. He joined ANSTO in 1999 where he managed nuclear fuel strategies and provided advice on nuclear issues to the Federal Government. He was subsequently appointed as Commissioning Reactor Manager and then the first Reactor Manager at ANSTO's OPAL research reactor. He is now a consultant and

Honorary Associate Professor and principal lecturer for Nuclear Reactors and the Nuclear Fuel Cycle at the Australian National University.

Commodore Denis Mole AM RAN Rtd

Denis Mole is a former senior naval officer and was Commander of the Submarine Force, having previously commanded individual submarines, a surface combat ship and a naval base. He was Deputy Director of Submarine Policy in Canberra when the Collins class submarines were ordered and he was the Submarine Operating Authority for Australia in Maritime HQ. After leaving the Navy he spent 12 years in senior executive positions in the commercial marine industry, whilst simultaneously assisting the navy as Commodore in the Reserves and as the submarine expert on the Chief of Navy's Seaworthiness Board.

Commodore Mark Sander RAN Rtd

Mark Sander is a former senior naval officer and submariner. He has been the Commanding Officer of submarines and the Deputy Commander of the Submarine Force. In Canberra he has served in both key submarine policy and project development roles. His latter naval service saw him working as the Director General for what was then the Future Submarine Program, now the Attack class submarine program. In addition to being the current President of the Submarine Institute of Australia, he is the Chief Operating Officer of Safran Electronics & Defense Australasia.

The Honourable Theo Theophanous

Theo Theophanous entered the Victorian Legislative Council in 1988 and served until 2006 as one of the two members for Jike Jika Province before the introduction of proportional representation. He served as a Minister in the Kirner Government and was leader of the opposition in the Legislative Council from 1993-99. From 2006-10 he represented the Northern Metropolitan Region and served as a minister in both the Bracks and Brumby Governments, holding portfolios in Energy, Resources, Industry, Trade, State Development and Major Projects. Since leaving politics, he has served on the Board of National Information Communication Technology Australia and the Metropolitan Planning Authority Board. He remains an active political and social commentator in major Australian newspapers.

Professor Stephen Wilson

Stephen Wilson is Professor at the University of Queensland in the School of Mechanical and Mining Engineering. At UQ he has a leadership role in the UQ Energy Initiative. He is also Managing Director of Cape Otway Associates, which provides commercial and policy advice, drawing on economic analysis and strategy in energy and resources for financial, corporate and government clients. Stephen was previously General Manager Market & Industry Analysis in the Energy Product Group at Rio Tinto.

Preface

Mark Sander
*President,
Submarine Institute of Australia*

*T*his book is unusual in that it does not represent the opinion of any one person. It has many authors, all possessing deep knowledge of their subject. While each chapter stands alone, when read together the contributions to this collection provide a picture greater than the sum of its parts. The Submarine Institute of Australia (SIA) and UNSW Canberra have combined to harness practical experience and technical expertise to start a conversation about nuclear energy and submarine propulsion. The SIA is the nation's leading organisation for all things relating to submarines. Its members are former submariners and industry experts who have devoted a large part of their professional lives to the place of underwater warfare in the configuration of Australia's defence and security provision. The SIA is dedicated to strengthening relationships between government and industry and, when necessary, offers advice on tactics, operations and strategy as well as logistics, maintenance and equipment. So, why this publish this book now and why a collaboration with UNSW? First, to the question of timing.

The 2016 Defence White Paper restated the intention of previous governments to double the size of Australia's submarine force from six to twelve boats, all conventionally powered, that is, diesel propelled. The renewed commitment was welcomed by those who considered the existing fleet was too small. At the SIA's annual conference in 2017, a number of retired senior submarine officers lamented that the Department of Defence had not given serious consideration to the prospect of the new submarines being nuclear powered. The implicit presumption was that informed Defence officials privately believed that Australia needed and should have nuclear powered submarines but recognised anything 'nuclear' was politically problematic. Put simply, merely suggesting that Australia might consider nuclear-powered submarines was largely pointless because national governments were unwilling to consume political capital on an issue that deeply divided the Australian people, and which would bring voter opprobrium at the next election.

Mere mention of the word *nuclear* sets off alarm bells for many people. Curiously, it would be much easier to discuss, 'fission generated, steam powered, turbine propulsion' than to discuss, 'nuclear power', although they mean the same. The word *nuclear* is clearly the problem. It generates fear and anxiety. Sadly, few in senior government positions have been willing to stare down uninformed commentary and provide intellectual rigour to a vital discussion that is closed down before gaining any momentum. This is, of course, the objective of many opponents of nuclear power and it is fundamentally anti-democratic and short-sighted. The lack of strong leadership has long blighted Australia's consideration of nuclear options for civilian and military needs with little attention to the national interest. I am not presuming that strong leadership will mean Australia establishes a nuclear industry but that, at least, a sensible discussion can take place and the advantages and benefits of nuclear power are properly judged against the costs and risks.

The discussion of nuclear powered submarines at our 2017 conference happened amid passionate debates about climate change problems and carbon emission policies. The senior submariners who contributed to the discussion were not conservative, right wing nor extremist. Their political sympathies were naturally mixed. Significantly, they were 'believers' in climate change; they had been convinced that climate change is bad and were persuaded that it resulted primarily from human activity, specifically carbon emissions. It was frustrating that nuclear power was not only being ignored as a means of propelling submarines, it was not even being considered as a means to reduce Australia's carbon emissions in a safe and cost-effective manner. If the world was experiencing a climate change crisis and Australia was in a position to exercise regional leadership, why not at least consider an option with considerable appeal? Given that so many of Australia's friends and allies relied on nuclear power for civilian and military applications, it made no sense that Australia was unwilling to examine a cost-benefit analysis.

Submariners are probably more aware of the benefits of nuclear power than the general public. Throughout the world, submarines represent the most extensive application of nuclear power other than domestic power stations generating electricity for industrial and household use. For the submarine community, it was the lack of public discussion regarding nuclear power that was difficult to comprehend in a technologically advanced, democratic country with the world's largest known reserves of uranium, the raw material required to produce nuclear power. There were many apparent benefits and few risks that could not be comprehensively mitigated. In turning incredulity into initiative, the senior submariners decided to provoke public discussion with no particular outcome in mind, other than identifying and advancing the national interest.

Now to my second question: why UNSW Canberra? To ensure the discussion was open, rigorous and balanced, the SIA established a collaborative partnership with UNSW Canberra. Drawing on its long and close relationship with Defence—UNSW has been the foremost provider of academic education to the Australian Defence Force since 1967—the University would contribute its expertise in the fields of national security and public administration. A joint SIA-UNSW seminar with the title, 'A Nuclear Industry Future for Australia? Starting the conversation', was held in October 2019 at the Australian Strategic Policy Institute (ASPI) in Canberra. The seminar was facilitated by Professor Tom Frame, a former naval officer whose more recent academic interests have focussed on the political-military interface. The seminar participants consisted mostly of scientists, engineers, academics, former politicians and submariners—current serving and retired.

A spirited keynote opening address was delivered by former Federal Labor Minister for Resources, the Honourable Martin Ferguson, who argued vigorously for his party colleagues to sponsor an even-handed consideration of nuclear power. Other valuable contributions were made by Professor Ken Baldwin (Australian National University), Professor Lyndon Edwards (Australian Nuclear Science Technology Organisation), Professor Ian Lowe (Griffith University), Professor Stephen Wilson (University of Queensland), Dr Ben Heard (Frazer Nash Consultancy), Peter Jennings (ASPI), Rear Admiral Steve Lloyd (former Royal Navy nuclear engineer and BAE nuclear ballistic missile program director), Tony Irwin (Engineers Australia), Robert Pritchard

(Energy Policy Institute of Australia), Katherine Ziesing (Editor, *Australian Defence Magazine*), Patrick Gibbons (Minerals Council of Australia), Greg Ward (Chief of Staff to South Australian Royal Commission into the Nuclear Fuel Cycle), Alan Behm (former senior public servant specialising in political and security risk), the Honourable Stephen Conroy (former Labor Minister for Communications and Sky News presenter), and the Honourable Theo Theophanous (former Labor Minister for Energy and Resources in Victoria).

The panel of speakers included those who were opposed as well as those who were supportive of nuclear power. Several of the speakers were ambivalent or just undecided. They could see the merits of both sides. Irrespective of their own position, there was a shared sense of frustration among speakers and attendees that previous discussion of nuclear power had been so unproductive. In fact, many despaired at the energy expended on pointless debates that had produced far more heat than light. Plainly, ignorance had flourished while insight floundered as wild assertions went unchallenged and reasonable objections were dismissed. It would be difficult to imagine an issue of such national significance in Australia's history that had been so poorly considered. The need for the seminar was obvious. The submariners were not alone in their despair. The opponents of nuclear power were certainly dismayed at the superficiality of those who shared their position but not their reasoning.

The 2019 seminar achieved its aim: a conversation has started and is continuing. A consensus has surprisingly emerged. It seems that Australians opposed to nuclear power stations are more likely to support the use of nuclear power for submarines. The reasons are not difficult to discern. Although many Australians are concerned about the cost and safety of power stations, they appear to be less anxious about compact reactors in submarines that operate far from their homes and business. The Government should take note of this changing mood as rising generations of young Australians with greater exposure to new technology and lesser exposure to old polemics want action on climate change and have become impatient with the slow progress of renewable energy solutions.

Is nuclear power still a viable option for domestic purposes? Nuclear power is still prohibited by legislation. The persistence of this largely symbolic prohibition is nonetheless stifling practical consideration of the risks and benefits of nuclear energy. The Federal Standing Committee on the Environment and

Energy conducted an inquiry into the prerequisites for an Australian nuclear energy late in 2019. Although the majority recommendations were generally in favour of developing nuclear power, dissenting opinions within the committee predictably emerged along partisan political lines. The familiar slogans and usual certainties of previous inquiries were gone but deep disagreement remains on both sides of the party divide.

Despite the Australian Government's understandable preoccupation with the global pandemic associated with COVID-19, it was sufficiently concerned with the deterioration of regional security since the publication of the 2016 Defence White Paper to release a Defence Strategic Update on 1 July 2020. It brought some clarity to the consideration of looming challenges but what will Australia's national security situation be in another 5, 10 or 15 years? As major Defence acquisitions involve decades of planning and development, what will be the prevailing conditions in 30 years when Australia is still building conventional diesel-powered submarines?

The aim of this book is ensuring we are asking not just the right questions but the *best* questions of how the Australian Government should protect its people and promote their interests. The SIA and UNSW Canberra hopes the chapters that follow will prompt policy analysts and political commentators to think more creatively and expansively about the future, and the possibility that nuclear power might enhance everyday living, promote the common good and dissolve tensions between nations. Information is power; insights are assets. This book will empower readers to contribute productively to a debate that matters to the entire planet and to every person.

Introduction

*A*ustralia has historically been an early adopter of new technology and Australians are usually receptive to practical innovation. Nuclear power is, however, a notable exception although it has been an option for civilian and military use for decades. Uranium was first discovered in 1789, closely followed by discovery of radiation and experiments on its potential use throughout the nineteenth century and into the twentieth century. The Second World War shifted research toward weapons development with the first atomic weapon tested in July 1945 and used the following month. The devastating consequences of deploying such weapons prompted the United States President, Dwight Eisenhower, to establish the 'Atoms for Peace' program in 1953. In the United States, development of nuclear power for both domestic energy supplies and for warship propulsion occurred tandem well into the 1960s.

Over the past 60 years there have been several Australian government technical inquiries and policy reviews into the merits of developing a local nuclear industry to generate electricity. Although most have recognised the potential benefits and possible advantages, and recommended public and private investment in exploring options, not even a single nuclear power station has proceeded beyond the concept stage, let alone been built. Of the G20 nations, Australia is one of only three that does not derive some of its electricity supply from nuclear power. Australia alone has a statutory prohibition on nuclear power. It is the only country with such a law and, ironically, it was drafted by a right-of-centre government that later commissioned a report into the viability of nuclear power for Australia's future energy needs. While the prohibition was prompted primarily by domestic political conditions, it would have an enduring influence on the provision of national security in a manner the Liberal-National Party Coalition did not contemplate.

In 2016, the Turnbull Coalition Government announced its commitment to build 12 French-designed *Attack* class submarines in Adelaide to replace the six ageing *Collins* class submarines. The new submarines were to be 'conventional' powered, meaning they would have diesel-electric propulsion. As the 2016 Defence White Paper claimed the 12 new submarines would be 'regionally superior submarines with a high degree of interoperability with the United States', it is curious that Australia had not previously considered nuclear power for either civil or naval needs. Simply observing that Australia does not have a nuclear industry is not an adequate answer. The principal problem is that whenever nuclear power is considered in Australia, the civil need for electricity generation and the security need for submarine propulsion have been considered separately. If considered holistically, the overall case for a nuclear industry might be stronger.

To test the hypothesis that Australia needs a nuclear industry that could support both civil and naval applications, the Submarine Institute of Australia (SIA), in conjunction with UNSW Canberra, hosted a public seminar in October 2019. The seminar's modest remit was 'A Nuclear Industry Future for Australia: Starting the Conversation'. This volume reflects the themes and issues that were canvassed by a series of presenters, many of whom have contributed chapters.

★ ★ ★ ★

For a country that does not host a local industry, Australia's nuclear power story is complex and controversial. To explain the present situation and interpret the prevailing mood, this book is divided into four parts. The problems and polemics associated with Australia's consideration of nuclear power—its generation and use—is explored in part one. In setting the scene, my chapter draws attention to the difference between *partisan* leadership and *public* leadership, and their influence on the nuclear debate in Australia. I neither argue for nor against nuclear power as I have no experience in drafting energy policy and no expertise in assessing competing technologies. But as someone with a background in the dynamics of political debate, I have examined the absence of public discussion and the presence of parliamentary posturing.

Neither has helped to bring any clarity to considering where the public interest might lie or how the public good can be furthered. Discussion of nuclear power is frequently emotional with reasoned arguments supported by evidence summarily dismissed with special pleading. As fear and not facts have tended to shape the discussion, it is hardly surprising that little practical progress has been made.

In chapter 2, Theo Theophanous, a former Labor parliamentarian and cabinet minister in Victoria, whose portfolio responsibilities included resources and energy, explains Australia's failure to achieve bipartisan energy policy in the context of managing carbon emissions and climate change. The absence of bipartisan goodwill has had a direct and defining influence on consideration of nuclear power. He helpfully explains some of the distortions that are often quoted in cost comparisons of various forms of electricity generation. Theophanous contends that national leadership, drawing principally on scientific facts and economic principles, need to place the public interest before partisan politics. There is, he argues, no alternative if Australians are serious about managing climate change at an acceptable cost.

In the quest for a reliable scientific basis for a public discussion, few people in Australia are more qualified to explain nuclear science and engineering than Lyndon Edwards of the Australian Nuclear Science and Technology Organisation (ANSTO). In chapter 3, Edwards provides a comprehensive overview of the last 70 years, including more than six decades of continuous nuclear reactor operations at Lucas Heights in Sydney. Noting that before 1987 the ANSTO and its predecessor organisations focussed their attention on nuclear power, Edwards explains the shift to nuclear science, production of medical isotopes and the technology needed to treat and manage nuclear waste. Importantly, he concludes with an observation that Australia has the capacity to support the design, procurement and implementation of small modular reactors either for energy production or naval propulsion.

Any account of Australia's nuclear history, capability and potential needs to include the existence and extraction of uranium given, its relative abundance in this country. In chapter 4, Mark Ho, President of the Australian Nuclear Association, explains that uranium was first encountered in Australia more than 125 years ago. Mining began in 1911 and, since 1954, has been extracted continuously. If read in isolation from political considerations, the natural assumption would be that uranium mining and nuclear power would feature extensively in Australia's national development. After all, this country has the world's largest known quantity of economically recoverable uranium reserves. It is only when read in the context of the earlier chapters that the lack of public leadership becomes more apparent. The consequences of the enduring lack of public leadership become clearer when we consider the breadth and depth of Australia's capacity for science and engineering. Put simply, Australia has always had the resources and the skills needed for a nuclear industry.

Nuclear technology and climate change

Part two features a discussion of the technology associated with nuclear power and its potential, partially to address one of the great dilemmas of contemporary policy and practice: how to reduce carbon emissions at an affordable cost. In chapter 5, Tony Irwin, Chair of the Nuclear Engineering Panel of Engineers Australia, offers a comprehensive explanation of the complete nuclear fuel cycle. Ben Heard describes the principal application of nuclear technology, electrical power generation, in chapter 6.

They note that opponents of nuclear power mostly assume their position based on apocryphal tales relating to safety or, more pointedly, the catastrophic risks associated with nuclear power. These opponents tend to dwell on three of the most notable nuclear incidents over the past 50 years: Three Mile Island in the United States (1979), Chernobyl in the Soviet Union (1986) and Fukushima in Japan (2011). There were no fatalities from nuclear radiation at Three Mile Island nor Fukushima. The incident at Chernobyl, while extremely serious and killing 31 plant operators and firefighters, has been consistently over-dramatised and its consequences exaggerated by advocates and journalists. The nuclear power station at Chernobyl was in the former Soviet Union. The reactor that exploded was a first generation Russian design of a type that would never have been approved in the West. A factual comparison with other forms of power generation and other industries, such as the chemical industry, would present nuclear power in a positive light. A 1984 disaster at the Union Carbide pesticide factory in Bhopal (Madhya Pradesh, India) is known to have killed more than 3700 people and injured half a million others. Reactor technology has advanced significantly since Chernobyl. In chapter Seven, Tony Irwin describes one of the most promising new types of generator, Small Modular Reactors (SMRs).

The real costs associated with generating sustainable electric power have been the subject of both unintentional and deliberate confusion. The total cost of coal power is rarely quoted as including the cost of damage to the environment or impact on health costs. Expenditure on renewables is rarely quoted as the cost without government subsidy, nor the cost of backup power for when the sun and wind fail to cooperate. Disposal costs of old or unserviceable hardware is rarely included in stated expenditures, especially the disposal cost for the thousands of rooftop solar panels. To navigate this labyrinth, Stephen Wilson, a senior academic at the University of Queensland, gives a factual account of the actual costs in chapter 8.

Plainly, nuclear power has a role to play in managing climate change. It is disappointing how frequently the loudest proponents of doing more to manage climate change are quick to depict their arguments on vague authorities, such 'the scientists say ...' or 'the United Nations concludes ...' They rarely acknowledge that thousands of scientists and a majority within the United Nations also recognise that nuclear power needs to be part of the world's energy solutions. In chapter 9, Ben Heard provides a good summary of the role that nuclear power is having and the potential for more positive impact on reducing carbon emissions.

But what has any of this to do with considering whether Australia should acquire nuclear submarines?

Nuclear technology and national security
In January 2020, the Australian National Audit Office (ANAO) conducted a review of the future submarine program. It was prompted by two considerations. First, the project involves expenditure of $A80 billion, making it the largest Defence procurement in Australia's history. The management of the project needs to be effective and efficient. Second, the decision not to acquire an off-the shelf submarine platform but to engage a 'strategic partner' to design and deliver the submarines with significant participation of local industry, has increased the risks.

What the ANAO review did not do was investigate whether there were any off-the-shelf platforms that could meet Australia's capability requirements for future submarines. Furthermore, noting the Government's requirement that the future submarines be 'regionally superior', the ANAO did not investigate whether or not the *Attack* class submarines will meet that criterion. It was probably assumed that they would. The contributors to part three of this book consider Australia's region of strategic interest and whether the future submarines meet the nation's needs. Although assessing the capabilities required in new submarines began with the 2009 Defence White Paper and were confirmed in the 2016 Defence White Paper, the strategic circumstances confronting Australia have changed considerably since then. Most commentators believe they have deteriorated, meaning they are less benign and less stable. Mindful of rapid political and economic change, the Government released the 2020 Defence Strategic Update and the 2020 Force Structure Plan although, surprisingly, neither document revealed any new thinking about submarines.

In contrast to the national security outlook in the 2016 White Paper which looked out to 2035, the 2020 Update acknowledged changes but remained silent on the more distant future. This was also surprising given the twelfth *Attack* class submarine will not enter service until about 2053 and is expected to remain in service until 2080—a span of 60 years. By comparison, none of the ships acquired by the newly established Royal Australian Navy in 1911 were operated for more than 20 years. After the Great War of 1914–18, naval planners were not required to look much beyond 1930. There was never any thought that they would be thinking about what might happen after 1980. The thought that defence and security needs can be projected across six decades with any confidence, let alone conviction, is either audacious or arrogant. In chapter 10, Malcolm Davis takes a more sanguine view in his account of recent developments regarding China and the potential influence the continued rise of the Peoples' Republic could have for Australia and its conventional submarine force.

Perhaps unhelpfully, the 2020 Update repeats the mantra of earlier White Papers that Australia's new submarines will be 'regionally superior' but offers little insight as to what that means. Denis Mole, one of Australia's most experienced submarine commanding officers, asserts in chapter 11 that conventional submarines are generally inferior to nuclear powered submarines, including those that China will be building in the 2040s and thereafter. Mindful of changes to the international laws of armed conflict since 1945, Mole asks whether the *Attack* class will be armed with the right weapons to deter adversaries and respond to threats. And if fitted with those weapons, can they be used effectively considering the manoeuvring limitations of conventionally powerer submarines? This is a critical question with a right and wrong answer.

Mole considers one more pressing question: how would Australia proceed if it decided to acquire nuclear powered submarines? He offers a series of options. The SIA supports proceeding with conventional propulsion for the first few *Attack* class submarines given it is probably too late for a nuclear-powered alternative to be afforded due consideration. But if Australia's strategic circumstances change, there ought to be scope for nuclear power to be considered for the later platforms. At a debate held at the National Press Club in March 2020, Peter Jennings from ASPI, Professor Hugh White from the Australian National University and Senator Rex Patrick from South Australia critiqued a report published by Insight Economics entitled, 'Australia's Future Submarine. Do we need a Plan B?'. None of the speakers thought the *Attack* Class should not be 'Plan A'. Jennings believed that Plan A must be made to work because it is simply too late for any alternative to be evaluated. White endorsed the Insight Economics recommendation that an 'evolved' *Collins* Class submarine—a 'Collins 2.0'—should be developed with government funding as a form of competitive pressure on Naval Group's *Attack* class. Patrick was less clear but appeared to support a Plan B, favouring a German 'off-the-shelf' option.

More significant were the comments of Jennings and White on nuclear power which seemed to converge. Jennings remarked in *The Strategist* in March 2020:

> In 2016, and still today, we do not have a realistic option to go for nuclear propulsion. Not without a decade-long investment to build the nuclear engineering, infrastructure, safety and operating experience the Navy would need.

In the same edition White observed:

In the longer term, there's real doubt that the *Attack* Class will do what we need. It's a very conservative diesel-electric boat with old-fashioned lead-acid batteries and it doesn't have air-independent propulsion. We may end up with a boat that can't meet the operational requirements that will be placed on it.

White went on:

So, the other key recommendation in the report is that, over the longer term, we look seriously at moving to nuclear propulsion. And if that is a serious option down the track, we need to start planning for it now, because the challenges involved are immense.

Readers should note the similarity between Jennings's assertion that before Australia could acquire nuclear submarines a decade-long investment would be necessary in nuclear related disciplines and White's argument that, given the challenges would be immense, planning ought to commence without delay. Although Jennings might prove to be correct, Australians will not know what might be required unless planning starts as White suggests. Notably, Jennings and White support proceeding with the conventional *Attack* Class program. Whereas White suggests nuclear powered submarines may be required in the longer term and we should therefore start planning now, Jennings opposes any distractions from building the *Attack* Class. The essential difference is not the future possibility of nuclear propulsion but whether need exists for a conventional powered Plan B submarine. Curiously, neither Jennings nor White explain why investigating nuclear propulsion should not form part of Plan B to provide, at the very least, competitive pressure.

Responding to the suggestion that Australia might be better served with smaller, cheaper, less capable, off-the-shelf boats rather than the large and expensive *Attack* Class submarines, Jennings retorted there would be little point in taking a Chihuahua to a Rottweiler convention. But the question to be answered in the 2040s: will the *Attack* Class submarines be the Chihuahua to China's Rottweiler? Will the *Attack* Class serve the nation's needs effectively in 20, 30 and 40 years' time or will a new generation submarine be required?

To those outside the Department of Defence, the future submarine project appears to be moving at virtual glacial speed. Little progress appears to have been made. The 2016 Defence White Paper foreshadows a review of future submarine technology in the late 2020s. According to the Government's Naval Shipbuilding Plan, the 12 new submarines are intended to be the first phase of a rolling acquisition program. The plan is for the continuous construction of naval vessels to sustain momentum, consolidate skills and achieve economies.

The SIA hopes that by stimulating conversation, the future submarine technology review will seriously consider nuclear power as a propulsion option for the submarines that will enter service during and after the 2040s. This collection is the start of a conversation we cannot afford not to have.

Part One

1 Nuclear power and the public interest

Tom Frame

Australians familiar with maritime matters know that Jervis Bay is integral to the nation's naval defence. It has also been the focus of several important public interest debates that have influenced the evolution of the Royal Australian Navy (RAN). The first of these debates was in 1901, shortly after Federation, arising from concerns that the New South Wales Government would have excessive influence over national affairs because of its then unrivalled economic power. To give the newly established national government access to the seas and a greater role in coastal and international trade, Jervis Bay was declared the national port and nearly 68 square kilometres of land was transferred from New South Wales to the Commonwealth. This was a very early attempt to re-configure Commonwealth-state relations for the benefit of all Australians. The RAN College at Jervis Bay was opened in 1915 with subsequent construction of operationally significant amenities and facilities, including the Jervis Bay Airfield, the Beecroft Weapons Range and the Shallow Water Sound Range, to support major fleet activities.

The second public interest debate was prompted by the Gorton Government's decision in September 1969 to consider construction of a 500-megawatt nuclear power plant at Murray's Beach on the southern shore of Jervis Bay. The plan did not survive the replacement of John Gorton, a former Minister for the Navy, as prime minister by William McMahon in March 1971. Based on the conclusion that coal-fired power stations were cheaper to build and easier to operate, the project was deferred in June 1971 and eventually cancelled by the Whitlam Government in 1973. This remains the first and only proposal for nuclear power to have received close consideration in Australian history. Its practical legacy, a high-quality access road and a large cleared area of land in the south-east corner of Jervis Bay, is well-known to most graduates of the RAN College. Its political legacy, that nuclear power is expensive and potentially injurious to the natural environment, has continued for just as long.

Introduction

There are few areas in Australia's national affairs that have suffered more from a lack of clear and consistent leadership than deliberations over possible provision of nuclear power. There has been some *political* leadership but very little *public* leadership, if the two are differentiated by public leadership's commitment to advancing the public interest. Political leaders have been consistently wary of raising the spectre of nuclear power. The reason is obvious. Atomic energy is a highly divisive issue within the community. It elicits emotional responses with unpredictable electoral consequences. Those energetically supporting the exploitation of nuclear power are countered by those vigorously opposing every aspect of the nuclear industry. Whether this support and opposition can be mobilised to effect voting behaviour is uncertain, leaving governments wary of raising the subject for public comment.

The paucity of public leadership is the focus of this essay. It has three parts. The first is an outline of what is meant by 'public leadership' and its relationship with the public interest. The second examines recent inquiries into nuclear power conducted by Federal and state governments, and the public interest considerations that have been nominated by both the proponents and opponents of nuclear power. The third identifies five factors that have complicated the application of a public interest test to nuclear power. I close by suggesting a two-stage process for a constructive discussion of whether Australia might consider acquisition of nuclear-powered submarines in the medium future.

Public leadership and the public interest

A consistent refrain in Australian history is the lack of good leadership. In his widely quoted work, *The Lucky Country*, published in 1964, Donald Horne described Australia as 'a lucky country run by second-rate people who share its luck'. Fifty years later, media commentators were lamenting debilitating instability within Australia's political leadership as the tensions between Tony Abbott and Malcolm Turnbull began to resemble the turmoil that enveloped the prime ministerships of Kevin Rudd and Julia Gillard. Was Australia incapable of producing competent men and women worthy of national leadership and capable of dealing with existing challenges and emerging opportunities?

The 'lack of good leadership' refrain can be interpreted as either a complaint about the political class or a lament that the country is deeply divided. It is convenient, of course, to attribute everything that is wrong with society to a handful of people and to their inability or refusal to give everyone what they want, when they want it. Just as likely a cause for discontent is the population's unwillingness to submit to leadership or their fickle reaction to the leaders producing compliance when it fulfils self-interest and defiance when it calls for personal sacrifice.

If asked for a brief description of what they seek in a leader, or if invited to identify the essence of leadership, I doubt that many Australians would have given their answers more than a second thought, let alone interrogated their own opinions for any bias or prejudice. Perhaps leadership is like art: people know what they like and what they do not like but they cannot quite explain their likes or their dislikes. Consequently, potential national leaders come and go, including among them viceroys, parliamentarians, heads of institutions, entrepreneurs, lawyers, academics, media personalities and sport stars until one captures the public's attention for reasons that may not be obvious. If their appeal is based on personality, as it often is, their influence might last a little longer before they say or do something effectively aligning them with an unpopular cause or an unfashionable mindset. Their standing then begins to decline.

The rise of identity politics, the suspicion of institutions and the spread of post-modernist angst have made it difficult for leaders to exert influence beyond the communities in which they themselves were nurtured and from which they initially acquired an authority to speak publicly. Attempts by leaders to refashion the nation or to reshape popular culture are resisted, if not resented, with the usual litany of objections ranging from the leader's failure to speak on behalf of every sub-group, to the leader's inability to understand the struggles of people who are unlike them or who might seek different things from life. Leadership is difficult in a society which seems to have perennially low regard for those who are elected or appointed to leadership positions. As the nation potentially dis-integrates into tribes and factions that are not coincident with the geographic boundaries of a state or local government area, the notion of public leadership becomes increasingly more complex as those things that divide gain greater prominence than those that unite.

Unlike the current fixation with what I would call popular leadership, *public* leadership is differentiated from other forms of leadership in that its focus is pursuing the public interest, a concept similar to, but distinct from, the common good and social capital. Public leadership does not seek to preference or prejudice the needs or wants of any group at the expense of others, nor does it advance personal or private aspirations and objectives to the detriment of the wellbeing of the whole population. Effective public leadership is indifferent to polemical agendas and partisan goals and transcends the practical preoccupations of administration and management. If the point and purpose of an activity is to further the interests (however these are understood) of private individuals or specific organisations rather than the entire nation, it should not be considered an exercise of public leadership.

Nuclear power involves a wide range of public interest issues whose balanced and conscientious consideration requires the exercise of firm public leadership to ensure the discussion of any proposal's merits does not drift beyond what will serve the interests of the Australian people, individually and collectively. Articulating these interests and clarifying their relative importance and priority are public leadership tasks.

Public interest and nuclear power

In contrast to its principal friends and allies, Australia has never relied on nuclear power as a foundation for national prosperity nor depended upon nuclear weapons as a bulwark of national security. As a country with a small population and a limited industrial base, economic modelling of nuclear power's benefits has proved a complicated exercise. As a well-connected medium power, there has been no pressing need for locally developed nuclear weapons although the advantages flowing from their possession have been canvassed from time-to-time. Australian governments in the 1950s and the 1960s occasionally considered the possibility of nuclear power and nuclear weapons but neither was pursued to acquisition.

In the early 1970s, an anti-nuclear movement gained momentum in Australia with candidates from the Nuclear Disarmament Party (NDP) elected to several parliaments in the 1980s, propelled by the Hawke Labor Government's support for uranium mining and endorsement of the United States' nuclear weapons program. Although the NDP was de-registered in 1992 and anti-nuclear rallies faded from view, it was not until the latter years of the Howard Government (2004–2007) that publicly declared attitudes toward nuclear power among legislators in Commonwealth and state parliaments began to change.

Conscious that its electoral appeal was steadily declining throughout 2005, the Coalition needed to address the effects of climate change and the impact of increasing energy costs. The 2006 report of the House of Representatives Standing Committee on Industry and Resources, *Australia's uranium— Greenhouse friendly fuel for an energy hungry world*, and the 2007 report of the Uranium Mining, Processing and Nuclear Energy Review Taskforce headed by former Telstra CEO, Ziggy Switkowski, provided impetus for a fresh consideration of nuclear energy in the context of mitigating Australia's greenhouse gas emissions.

Both inquiries recognised that, first, every country with an economy larger than Australia's relied on nuclear energy; and, second, Australia was the only country among the world's 25 leading economies that had excluded nuclear energy from its baseload power supply. While the Standing Committee's report could be dismissed as a partisan document reflecting the Coalition's position on nuclear energy, the report of the Switkowski-led taskforce, consisting of well-regarded scientists and academics without political affiliations, was disparaged by some critics as little more than a 'roadmap for Australia to go nuclear'. This was unfair and unjustified commentary. Precluded by its initial brief from making formal recommendations, the taskforce thought 'nuclear should be on the table' as it provided an important option for policy-makers to consider. The Howard Government was emboldened to act by both the report and its reception.

In opening a new research reactor at Lucas Heights late in April 2007, the Prime Minister, John Howard, spoke of nuclear power as 'a source of hope' that ought to be 'part of Australia's future'.[1] The Government declared its intention to launch a comprehensive nuclear power program with the prospect of more than 50 generation plants being built throughout the country. In response to the Howard Government's newly acquired enthusiasm for nuclear power, the Tasmanian Labor Government led by Paul Lennon and the Queensland Labor Government led by Peter Beattie considered legislation to ban nuclear power generation. The aim was ending any discussion of nuclear power before a debate had even begun. A representative of the Queensland Nuclear Free Alliance, Robin Taubenfeld, was surprised by the Howard Government's 'highly irresponsible' policy announcement which she described as 'political suicide'. As there was a Labor government in every state and territory, the Federal Coalition explored a series of legal routes to override state objections should their governments stand in the way of Canberra's plans. This was a less than satisfactory basis on which to conduct a discussion that required the goodwill of parliamentarians and the people throughout the country. In this context, discerning the public interest had already been displaced as the foremost consideration for legislators.

The Federal Labor Opposition led by Kevin Rudd campaigned strongly against nuclear power ahead of the 2007 election. Rudd described the Howard Government's plans as being 'too expensive, too dangerous, too slow, when it comes to impact on greenhouse gas emissions'.[2] During the campaign a number of Liberal candidates sensed the electoral unpopularity of the Coalition's support for nuclear power and distanced themselves from the policy. After nearly 12 years in power and the electorate eager for change, the Coalition was defeated at the November 2007 election with a large swing to Labor. But within 12 months of taking office, there were signs the Rudd Government was more open to nuclear power than the electorate had been led to believe. Switkowski thought Labor could be persuaded as to its merits and told a business forum:

> We will get there. I'm sure we will get there, whether it happens in the next term of government or the one after ... The attitude in Australia, I think, will move from concern to grudging acceptance, to enormous relief that we have this very efficient technology and these vast reserves that will give us ... the lowest cost, safest, cleanest form of base load electricity.

Notwithstanding Switkowski's optimism, Federal Labor maintained its opposition to expanding the nuclear industry although the Gillard Government eventually agreed to sell Australian uranium to India and senior ministers, Martin Ferguson and Gary Gray, were known to support closer consideration of nuclear power. Elsewhere, Labor's stance was inconsistent.

In March 2015, the Weatherill Government in South Australia announced it would hold a royal commission into the Nuclear Fuel Cycle. The inquiry would be conducted by former state governor and retired naval officer, Rear Admiral Kevin Scarce. Despite being accused of personal bias towards the nuclear power industry when appointed, Scarce asserted his openness to all points of view. This openness was implicit in the findings and recommendations of the report which was completed in May 2016. The Royal Commission supported establishment of a nuclear waste facility and expansion of mining and export activities. While the processing of uranium and the generation of nuclear power were unlikely to be financially viable within South Australia, Scarce recommended repealing Federal and state prohibitions on expanding the nation's nuclear industry. By the end of 2016, however, the Weatherill Government appeared to have lost interest in the report it had commissioned. A change of government following the March 2018 state election did not lead to a revival of interest in either the inquiry's findings or recommendations. The new premier, Steve Marshall, said nuclear power was not on his 'short term' radar although it could 'come back on the agenda down the track' if needed to achieve cheaper electricity prices.

Disappointed his report was effectively 'shelved', Scarce remarked in July 2019 that:

> we have to find a way to restart this discussion. Whether the answer ends up being for or against, you can trust the Australian people to have a mature discussion. At the moment we can't even have the discussion because it's not politically acceptable. The longer we go without having the debate, the less options we have for the future. We are not giving ourselves the options we need and we only have a limited time to do so.[3]

His judgment that the Australian people could be trusted to participate in a mature debate curiously overlooked the 'verdict' of the 'citizens' jury' convened by the Weatherill Government to consider his recommendations.

This form of 'deliberative democracy' was widely criticised as a flawed approach for assessing complex public interest claims,[4] but earlier expressions of bi-partisan support rapidly evaporated when the jury rejected Scarce's recommendations citing a 'lack of trust' in regulators and regulations. Lamenting that 'the citizens' jury cut the legs out from under [his report]', Scarce wanted discussion to continue. Had it continued, he remarked, 'we would have had an answer one way or another, and at the moment we don't'. He pointed to Australia's 'terrific nuclear record' and the capacity to 'develop a regulatory system that would be the best in the world'. Plainly, he was in favour of expanding Australia's nuclear industry. There remained, however, a requirement for 'the social will to do it, and that's the thing that worries the politicians'. In essence, could the public overcome its anxiety if a policy were shown to serve their interests?

The challenge for those who were undecided within the parliament and among the people was sifting the available evidence and evaluating competing claims when discussion of nuclear power was highly emotive and obviously polarised. Those in favour of nuclear power characterised their opponents as 'primitivists' living in the distant past and 'alarmists' who exploited ignorance to exaggerate community fear of catastrophic accidents resembling the systems failures at Three Mile Island in 1979 and Chernobyl in 1986, and, following a natural disaster, such as at Fukushima in 2011.

Some anti-nuclear activists portray their opponents as people so far to the right of the political spectrum that even right-wing ideologues think they are right wing ideologues. In an article claiming the nuclear power debate had become the new frontline in the nation's culture wars, a spokesman for Friends of the Earth Australia, Jim Green, claimed the 'far-right supports nuclear power if only because the "green left" opposes it'. Green asserted that 'support for nuclear power is increasingly marginalised to the far-right. Indeed support for nuclear power has become a sign of tribal loyalty: you support nuclear power (and coal) or you're a cultural Marxist, and you oppose renewables and climate change action or you're a cultural Marxist'. In portraying supporters of nuclear power as an 'extremist' cabal, Green included the Minerals Council of Australia, the Institute of Public Affairs and *The Australian* newspaper as part of a concerted attempt to 'wedge' the Labor Party and the Green Left. He quoted Greens Senator Sarah Hanson-Young:

Talk of overturning the ban on nuclear power in Australia is crackpot stuff. Aside from being a dangerous technology, nuclear power is wildly expensive and would take a decade or more to build. It would be a funny joke if it wasn't so embarrassing to have the Nationals, who are in government and who sit around the cabinet table, pushing for this. These people are meant to be in charge, and they're running around like a bunch of lunatic cowboys.

Intentionally polarising the debate was not restricted to activists and politicians. In January 2019, the Climate Council, which consisted mainly of academics and former civil servants, issued a media release contending that nuclear power plants:

> are not appropriate for Australia—and probably never will be. Nuclear power stations are highly controversial, can't be built under existing law in any Australian state or territory, are a more expensive source of power than renewable energy, and present significant challenges in terms of the storage and transport of nuclear waste, and use of water.

The statement was notable, first, for use of the word, 'appropriate'. It was a curious choice implying conflicting values rather than objective appraisal of cost and benefit. Second, was the observation that nuclear power plants were 'controversial', implying that being controversial *ipso facto* made them inadvisable.

The Climate Council was formed by members of the Climate Commission, a body established and funded by the Gillard Government but later disbanded by the Abbott Government on grounds of administrative efficiency (the Council's functions were transferred to the Department of Environment). The Council has been accused of polemical bias but stands by its independence and autonomy as a non-for-profit organisation funded by community dona-tions. Perhaps unwisely given its apolitical aspirations, its CEO lamented the re-election of the Morrison Government in May 2019 and applauded the failure of Tony Abbott, the Coalition's foremost climate change sceptic, to retain his seat in the House of Representatives. Comments of this kind cast the Coalition parties as adversaries to be defeated rather than public leaders to be persuaded.

The unexpected Coalition election victory in May 2019 revealed that opinion polling had become a fraught activity producing unreliable 'results'. Most pollsters had tipped a substantial swing against the Coalition and the election of a Labor government, citing environmental concerns as an important vote-winner. Among the lessons to be learned from the polls' failure to depict the electorate's mood accurately, it became clear that what an individual voter thinks about a particular issue considered in isolation will not necessarily determine how he or she will cast their vote at the ballot box. There are many and diverse reasons shaping a voter's decision to preference one party or one candidate above and before another. There are also far fewer 'rusted on' voters. The once predictable tribal loyalties associated with Australian politics are rapidly dissolving. With more 'undecided' voters than ever before, the major parties are experiencing rapid and substantial swings for and against both their candidates and their policies. It also seems that growing concerns about climate change and energy security have made voters less likely to punish parties willing to put all options 'on the table', including the nuclear option, to contain energy costs and guarantee energy supply.

Evidence of greater community readiness to considering nuclear power was behind the inquiry into the nuclear fuel cycle established in August 2019 and conducted by the House of Representatives Standing Committee on the Environment and Energy. It addressed a number of issues: waste management and storage; health and safety; environmental impacts; energy affordability and reliability; economic feasibility; community engagement; workforce capability; security implications; and, national consensus. Openness to considering nuclear power should not, of course, be mistaken for endorsing the construction of reactors. Considering is not deciding. Nonetheless, the inquiry chair, Liberal member for Fairfax, Ted O'Brien, thought concerns about climate change and advances in nuclear technology were 'game changers': 'you can't contend there's an existential threat to life as we know it due to climate change and then oppose the cleanest form of industrial-scale energy generation the world has seen.'[5] He was optimistic that 'we have a national debate ahead of us on a major energy issue—nuclear, no less—without the hysterics and hyperbole that has dogged energy policy in this country for far too long.'

O'Brien's words were echoed by the New South Wales Treasurer, Dominic Perrottet, who deemed energy security 'the biggest challenge of our time'. He thinks Federal legislators should consider nuclear power 'and not just putting it off the table for ideological reasons from the past'.[6] The state's One Nation leader, Mark Latham, shared Perrottet's concern that the economy rested precariously on energy security. But this concern does not mean the public will, or necessarily should, support nuclear power.

The committee's report, *Not Without Your Approval: A Way Forward for Nuclear Technology in Australia*, appeared in December 2019 and made three recommendations.

> First, that [the Government] consider the prospect of nuclear tech-
> nology as part of its future energy mix; secondly, that it undertake a
> body of work to progress the understanding of nuclear technology in
> the Australian context; and thirdly, that it consider lifting the current
> moratorium on nuclear energy partially—that is, for new and emerging
> nuclear technologies only, and conditionally—that is, subject to the
> results of a technology assessment and to a commitment to community
> consent for approving nuclear facilities.[7]

The Labor members of the committee submitted a dissenting report. They recognised that for

> Australia to change the long-held bipartisan position against the
> development of a nuclear power industry in Australia it would have
> to make sense to do so. Yet on any analysis it doesn't, as the evidence
> to this inquiry has shown. Above all, there is no economic case for
> pursuing nuclear energy.[8]

The dissenting report noted that

> events (like Fukushima), innovations and advances in renewable energy,
> and emerging climate and energy system developments of the last ten
> years have made nuclear power even less relevant and appropriate in the
> Australian context at a time when nuclear power is already in decline
> elsewhere. There is simply no case for wasting time and resources on a
> technology that is literally the slowest, most expensive, most dangerous,
> and least flexible form of new power generation.

In the context of the key considerations

> namely, our future energy needs, the changing nature of our energy system, the comparative costs and delivery timelines of different sources of generation, the serious risks and dangers to the environment and public health, and the impact in terms of regional nuclear proliferation— the pursuit of nuclear energy in Australia would be deeply irrational.

The Independent member for Warringah, Zali Steggall, supported the first two recommendations in the majority report and acknowledged

> that an independent community engagement program should educate and inform Australians on all energy technologies including nuclear. I do not support recommendation 3 [removing the nuclear ban], which seeks conditional removal of the moratorium on some nuclear technologies. The Committee adopting recommendation 3 is pre-emptive. Obtaining a social licence is an essential prerequisite to any consideration of raising the moratorium on nuclear energy.[9]

In a separate public statement, committee chairman Ted O'Brien stressed that if Australians were 'serious about reducing greenhouse gas emissions, we can't simply ignore this zero-emissions baseload technology. Australia should say a definite 'no' to old nuclear technologies but a conditional 'yes' to new and emerging technologies such as small modular reactors'.[10] In reply, the Minister for Energy, Angus Taylor, stated the government had 'no plans' to lift the moratorium, affirming that 'any changes to the moratorium would need bipartisan support and broad community acceptance'.[11] A spokesman for the Australian Conservation Foundation, David Sweeney, said lifting the ban would start a 'conga line of supplicants to Canberra promising low carbon energy and seeking high public subsidy'.[12] The committee's report highlighted the key consideration: the public want to be informed before it gives consent.

There remain many hurdles to serious consideration of nuclear power. The public is still concerned about cost and safety although community acceptance of nuclear power rose slightly according to an Essential poll published in June 2019. This acceptance was tainted by self-interest. In response to the question: 'would you be comfortable living close to a nuclear power plant',

60 per cent of respondents said 'no'. Only 28 per cent of respondents said they were unconcerned.

Residual anxieties continue to arouse political opposition with claims of distortion and even deceit on the part of nuclear advocates. My UNSW Canberra colleague, Associate Professor Heiko Timmers, is an experimental physicist. He thinks the cost of replacing coal-generated electricity with renewables 'could be huge. These costs may *possibly* exceed those of building nuclear power stations'.[13] While the business case for nuclear power was 'shaky', he thinks 'a much stronger argument can be made for the back end of the nuclear fuel cycle: storing nuclear waste'. In reply, Paul Richards, an American seismologist, claims that storing nuclear waste was a 'trojan horse' that obscured a desire to introduce nuclear power and nuclear weapons.[14] In essence, Richards argues, any expansion of the nuclear industry will increase the possibility of nuclear weapons proliferation. Is this a reasonable observation or an questionable tactic?

Notably, 30 years after his government looked seriously at nuclear power, John Gorton told a Sydney newspaper that 'we were interested in this thing because it could provide electricity to everybody and … if you decided later on, it could make an atomic bomb'. The Gorton Government signed the nuclear non-proliferation treaty the following year (1970) but the prime minister had no intention of ratifying it, a task which was left to the Whitlam Government. Gorton's foremost advisor on nuclear policy was the first Vice-Chancellor of the University of New South Wales, Professor Sir Philip Baxter. A chemical engineer and a vigorous advocate of nuclear power, Baxter chaired the Australian Atomic Energy Commission from 1957 until 1972. He also believed the advent of nuclear power had military applications, declaring in 1975:

> Over the years I have initially advocated that we should create the necessary technology and industrial background to enable us to move into a nuclear armament quickly. More recently things have changed internationally. I'm now of the opinion that we should begin actively to create nuclear weapons for the defence of Australia.[15]

Baxter was an unashamed intellectual elitist. He argued that in the provision of energy, 'only experts understand the problems and can advise governments on what they should do'.[16] He cited the analogy of aircraft pilots: because public safety depends on them being experts, only those who understand the possibilities and potential of nuclear power can advise governments on energy policy and provision. He was largely unconcerned with the political, economic, social and ethical issues associated with nuclear power. Baxter fully supported the Gorton Government's Jervis Bay proposal but was ill-equipped to consider the environmental objections to its location which would only have increased with time. The site's most significant virtue, that it was owned by the Commonwealth, meant state government approval to proceed with the project was not required. Baxter knew that developing nuclear weapons was a much higher hurdle politically than providing nuclear power. He was ultimately unable to present a politically acceptable case for either.

To counteract continuing suspicion that the advent of nuclear power will increase the likelihood of Australia developing nuclear weapons, the case for nuclear power, and nuclear-powered submarines in particular, will need to assure the electorate that nuclear weapons are not, and will not, be part of an expanded nuclear industry in Australia.

A way ahead

There are five factors that could complicate the application of a public interest consideration of the issues associated with nuclear power. They can be briefly summarised: a short-term electoral cycle; a fractured political system; conflicted Commonwealth-state relations; a partisan media; and a self-interested electorate.

Recent Australian governments have tended to be in permanent campaigning mode. With the period between elections being no more than three years, the party winning office has a window of between 24 and 30 months to implement its policies before facing the people. As Australian electorates usually give first-term governments a second term (the last one-term administration was the Scullin Labor Government which was elected in 1929 and defeated in 1931), the party forming government can usually rely on five to six years in power. Even so, few seem to have an agenda extending beyond one term (noting that only three administrations since Federation have actually used their entitlement to a full three-year term). As any decision on nuclear power will require substantial investment capital and will not produce any community benefit for at least 15 years, the party making the decision to proceed with nuclear power will quite probably have to absorb all the political pain without experiencing any political gain during its term in office. Such a decision would need to be politically selfless.

Australia's adversarial political system, which sets a government against an opposition, can obscure as much as clarify the public interest. The two-party system has a role in testing and tempering the assumptions and claims associated with government policy development and decision-making. It can also divert attention from close consideration of the public interest when an opposition deliberately takes a contrary view—irrespective of whether it can be reconciled with a rationally considered assessment of the public interest—for the purposes of denying the government supremacy in parliament and credibility in the electorate. As all parties are committed to broadening their support base and neutralising their opposition, the government is also susceptible to choosing the politically most palatable policy option and the least confrontational administrative action instead of committing itself to the policy and the decision that most effectively and efficiently advances the public interest. While politics is essentially a contest of ideas, willingness to engage in compromise for political purposes might assist the ruling party to make a decision but comes at the expense of a single-minded pursuit of the public's interests.

The evolution of Commonwealth-state relations also works against an even-handed consideration of nuclear power. Since Federation in 1901, the Commonwealth has expanded its capacity to govern because of its enlarged revenue raising capacity. After 1910, when the Commonwealth was no longer obliged to return three-quarters of customs revenues to the states and 1942 when the High Court ruled that Commonwealth income tax had priority over state income taxes, the Commonwealth has regularly used its financial power to impose its political will on the states. The limits of the Commonwealth's legal powers and the states' rights to resist the exercise of those powers have been the subject of several High Court cases concerning the Commonwealth's desire to act in what it considers the national interest and the determination of a state government to oppose such action within its territory. In several instances, a state government has opposed the exercise of Commonwealth power to demonstrate its commitment in defending the rights the state and the interests of its people. Conversely, the Commonwealth has opposed the decision of a state government which it deems contrary to the national interest or a violation of Australia's international obligations. A Commonwealth decision relating to nuclear power would require action on the part of a state government, action which could be thwarted by a power struggle between one or more jurisdictions especially in circumstances in which a decisive electoral advantage was apparent.

The media are vital to the flow of public information and crucial to interpretations of its significance. Aside from organisations and publications that represent a particular segment of the electorate or appeal to a certain body of opinion, the mainstream media—television, radio and newspapers—continue to 'frame' conversations about matters of public interest. Aside from concerns about the influence of media ownership on editorial policy, the narrowing of opinion and the increase in polemical commentary is likely to shape public discussion of nuclear power and the potential exploitation of this issue for partisan purposes that may not serve to elucidate the relevant public interests.

Finally, a popular culture that emphasises individual wants ahead of collective needs can distort the consideration of the public interest by elected and appointed officials. Long-term investments frequently have long-term consequences. The decision to expand Australia's nuclear industry will require investment in infrastructure and impose a continuing burden on the nation's finances. One generation may be unwilling to fund benefits accruing to the next. Future generations may lament the debts left by their forebears. Conversely, the cost of research into renewable energy development might ultimately exceed expenditure on nuclear power but ultimately prove better for the natural environment. In such circumstances, electricity prices might continue to rise but the nation is freed from the burden of nuclear waste management. Decisions about public investment in energy sector research and development involve more than economic imperatives notwithstanding that a self-interested electorate might be drawn to a solution that serves individual financial interests narrowly conceived more than public social interests broadly considered.

Given the existence of so many complex issues, a two-stage process would assist a public interest consideration of nuclear power. The first stage is identifying all of the relevant considerations that bear upon the public interest and determining, on the basis of those considerations, whether the public's interests are advanced by nuclear power. The current inquiry has identified some but not all of the relevant public interest considerations. The second is contingent on the first. If a public interest case can be made for nuclear power, as much of the public as possible need to be convinced that nuclear power indeed serves their interests and that consensus is needed to support the immediate costs incurred to pursue long-term benefits.

If the case for nuclear power cannot survive a rigorous public interest examination, the matter is settled. But if a case can be made, the government will need to embark on a journey with the Australian people. Their support is contingent on gaining and maintaining their trust, an important point made by Michael Angwin, former chief executive of the Australian Uranium Association. He argued that trust is indispensable to the establishment of an Australian nuclear industry and building trust, not power-generating reactors, is the foremost task for its advocates in the first instance. Angwin thinks that fears of nuclear power are rational but that trust is the 'antidote to fear'. Such trust is not the outcome of promises but actions. He suggests that 'bipartisan political support is one of the starting points for building trust'. Such trust would be built on a series of commitments which he suggests might include:

We will not create the conditions for a nuclear industry unless Australians trust it and support it.

We will not proceed from one stage to another unless there is trust and support.

We will not force any nuclear facility on any community.

Any nuclear development must not foreclose on any other community, industry, economic or infrastructure development.

Such an approach will

enable people to make up their own minds. Clearly, this approach anticipates that people may well make up their minds to oppose nuclear power. Those with a strong belief in the energy and climate change benefits of nuclear power may find this a challenging possibility. But unless the trust-building task is approached in that way, it will likely fail. The overarching demeanour of the government that embarks upon this path should be sceptical, disinterested and open-minded.

Angwin's contention is built on broad agreement that nuclear power advances the public interest. It assumes the exercise of public leadership.

The challenge of public leadership in this context is two-fold. The first is creating an environment in which individuals are encouraged to overcome their own self-interest in the expectation that the public of which they are members will be enriched by the collective pursuit of interests that are common to all. The second is building public confidence in the handling of evidence and the sifting of argument to identify the most efficient and effective means of fulfilling a shared aspiration—cheaper and more reliable energy. Both of these tasks transcend the formal remit of parliamentarians and administrators. Organisations like the Submarine Institute of Australia are essential to hosting a wide-ranging discussion and encouraging a genuine contest of ideas. In the absence of community goodwill and consensus that the public interest is being served, any policy is liable to provoke resentment and resistance. Without individuals in the community believing their destiny is better served by working with instead of against others, the public interest will be little more than an empty political slogan. Leaders are key players and their leadership is decisive.

Endnotes

1   Richard Macey, 'Nation's energy future is nuclear: Howard', *Sydney Morning Herald*, 21 April 2007, p. 1.

2   Katharine Murphy, 'PM puts faith in nuclear power', *Age*, 30 December 2006, p. 1.

3   David Penberthy, 'It's time to push nuclear option', *Australian*, 25 July 2019.

4   https://indaily.com.au/opinion/2016/11/11/inside-the-chaos-and-bias-of-the-citizens-jury/

5   https://www.theaustralian.com.au/commentary/nuclear-debate-without-hysterics/news-story/d8c0fb92c54697b1aa439ca8f3dc34b8

6   https://www.smh.com.au/politics/nsw/biggest-challenge-of-our-time-perrottet-s-warning-on-energy-security-20190829-p52m6c.html

7   Foreword, https://apo.org.au/sites/default/files/resource-files/2019–12/apo-nid271551.pdf, p. 5.

8   https://apo.org.au/sites/default/files/resource-files/2019–12/apo-nid271551.pdf, p. 55.

9   https://apo.org.au/sites/default/files/resource-files/2019–12/apo-nid271551.pdf, p. 76.

10 https://www.tedobrien.com.au/article/nuclear-energy-not-without-your-approval, see also : https://www.tedobrien.com.au/sites/default/files/2020–02/200205%20Environment%20and%20Energy%20Committee%20Report.pdf

11 https://www.smh.com.au/politics/federal/taylor-rejects-call-for-partial-lift-of-nuclear-power-ban-20191213-p53jsf.html
12 https://www.smh.com.au/politics/federal/taylor-rejects-call-for-partial-lift-of-nuclear-power-ban-20191213-p53jsf.html
13 https://theconversation.com/australia-should-explore-nuclear-waste-before-we-try-domestic-nuclear-power-121361
14 See https://antinuclear.net/2019/08/06/paul-richards-refutes-heiko-timmers-push-for-australia-to-import-nuclear-wastes/
15 Christine M Leah, *Australia and the Bomb*, Palgrave MacMillan, New York, 2004. p. 83.
16 Philip Baxter, *Canberra Times*, 27 June 1979.

2 Politics versus rationality

Theo Theophanous

The rough and tumble of politics has shaped the energy debate in Australia for decades, including whether nuclear energy should be considered an option. Opposition to nuclear power and nuclear armaments is a feature of post-war politics that cemented itself in the public consciousness of Australians and has been supported by both sides of politics. It included periods of intense opposition, such as when the French were conducting atmospheric nuclear tests in the Pacific the 1960s and 1970s, and other periods during what became known as the 'arms race'. Despite Australia's enduring opposition, nuclear weapons proliferation has occurred to a number of countries in our region including China, Pakistan, North Korea and India. This remains a challenge no matter how much Australia spends on conventional weapons.

Differentiating atomic bombs from nuclear energy, supporting the latter while opposing the former, requires a thoroughgoing deconstruction of the anti-nuclear ideology—theology. I use the term *theology* because the opposition to nuclear in all its forms is pursued by some with the kind of religious zeal that transcends logic or science or even politics. Some political leaders have sought to revisit nuclear energy objectively, concentrating on the logic and the facts and, most importantly, the morality of Australia's position. Prime Minister Bob Hawke pointed out in 1984 the irrationality and absurdity of allowing uranium mining at the Ranger and Narbalek mines in the Northern Territory but not at the Roxby Downs mine in South Australia. It was the power of this logic that led to the 'three-mine-policy'. Other decisions to expand the industry were made, such as allowing uranium sales to France in the 1980s although it did not stop all nuclear testing in the Pacific until 1996, and to India despite its not signing the nuclear non-proliferation treaty.

Today, with a third of the world's uranium reserves, huge extraction and export operations, and more mining than ever before, Australia is entangled in the world nuclear cycle. Australian exports of yellowcake (a uranium concentrate U308 produced using acid or alkaline leach technology) now averages 8600 tons annually generating significant revenues, as well as jobs in regional areas, taxes and royalties.

For Bob Hawke, Australia was complicit in the entire nuclear cycle and should do more. In 1991, he commissioned a report from the then Chief Scientist, Ralph Slatyer, to ascertain whether Australia should accept some greater responsibility for its complicity by offering to store the nuclear material it sold to other countries. The report found that 'Australia has the safest remote geological formations in the world for this purpose'. As late as 2014, speaking at the Indigenous Garma festival in the Northern Territory, Hawke proposed using royalties from nuclear storage to benefit Indigenous Australians: 'We need to do something substantial to finally eliminate these disgraceful gaps in well-being and lifetime opportunities'. Notwithstanding support from Adam Giles, the Northern Territory Chief Minister, and the Northern Land Council, the proposal was abandoned. Another opportunity was lost.

Jay Weatherill, the Labor Premier of South Australia, initiated a royal commission into the storage of nuclear waste. It seemed a logical initiative given South Australia's increasingly difficult energy security issues and the export of uranium from that state. The commission concluded in 2016 that a nuclear storage industry was viable. Without bipartisan and national political support, this idea also went nowhere.

Since Harold Holt's time as prime minister (1966–67), a succession of Liberal leaders have also considered the appeal of nuclear power. More recently, in 2007, John Howard commissioned an in-depth look at nuclear power pricing led by the former CEO of Telstra and nuclear physicist, Ziggy Switkowski. he found that nuclear power would cost between 20–50 per cent more than coal or gas but was cheaper than both if an emissions trading scheme was introduced that imposed a price on carbon. As an emissions trading scheme was never going to happen while John Howard was prime minister, Switkowski effectively killed the nuclear option with his finding that it would be uncompetitive without an ETS.

Unions, too, have been divided politically over the question of nuclear power. The Australian Workers' Union and the mining division of the CFMMEU have supported nuclear power but many other unions remain opposed. Recently Daniel Walton, the Australian Workers' Union national secretary, has argued that it is 'ludicrous' for the progressive side of politics to continue to reject 'zero emissions technology' when 31 advanced countries are relying extensively on nuclear power for their domestic needs.

Australia itself has a nuclear reactor. The Lucas Heights facility in Sydney commenced operations in 1958. The initial reactor was replaced by a new Open Pool Australian Light water (OPAL) reactor in 2007. Australia's new reactor is one of a small number of reactors worldwide that produces millions of doses of much-needed medical radioisotopes for research and direct use. It also conducts scientific analyses for the mining sector and other industries. Despite the existence of OPAL and good work conducted by the Australian Nuclear Science Technology Organisation (ANSTO), the successor body to the Australian Atomic Energy Commission, the bipartisan political wall of resistance to nuclear energy remains unmoved.

Modern political energy wars

Whenever, in the past, politicians have attempted to 'float' the nuclear option, the debate has been subsumed by intense energy policy differences and ideological energy wars within and between political parties. These wars have claimed four prime ministers as the Left and Right factions within the major parties pursued policies that were later modified or dumped with the change of leader. The energy wars were not purely, or even mainly, about nuclear power. They were and are about our energy mix, and specifically how we deal with climate change by reducing carbon emissions. Before considering the politics associated with nuclear energy, we should first explore the politics involved in the existing and proposed conventional energy solutions.

When I was the Minister for Energy Industries in Victoria (2002 to 2006), virtually all political parties agreed that we could not transition from a coal-fired economy to 100 per cent renewable energy without a substantial increase in the use of natural gas during the transition period. This position was held by the Left and Right of the Labor Party and implicitly sanctioned by the Greens and the Coalition parties. After all, this was precisely the settlement that had been negotiated in Europe, particularly in the United Kingdom, as it transitioned from coal to gas using North Sea gas reserves to produce electricity and to retire dirty coal fired power stations that had been operating for decades.

In the 2000 Victorian state election campaign, the Greens changed their political positioning to oppose all forms of fossil fuel use, including gas, in power production. They began arguing that Australia could transition directly from 82 per cent fossil fuel provided energy, 71 per cent of which was coal, to 100 per cent renewable energy. The major political parties, in particular the Labor Party, allowed itself to engage in this new narrative and ultimately to succumb to it. Political leaders gave tentative approval to Greens' claims that renewables could produce all the power we needed with backup power to be provided by pumped hydro, batteries or, more exotically, hydrogen fuel cells. The only disagreement was about time frames: would this happen in 2030, 2040 or 2050. Nuclear was off the table for all parties in this unfolding scenario.

The energy wars not only ended political careers, they also buried many good policies designed to make the transition to renewable energy more possible. These policies included the Emissions Trading Scheme on the Labor side and the National Energy Guarantee on the Coalition side.

The 'technology roadmap' and bipartisanship

The Morrison Government's Energy Minister, Angus Taylor, has proposed moving past ideology to science with his vision of transitioning (eventually) to a low carbon future through a 'Technology Roadmap'. Taylor says in a ministerial foreword to the plan, 'this is about technology not taxes … It is an approach based on rigour, discipline and optimism, not ideology'.

These are fine words but the important political message is about taxes. The policy deliberately lacks a target because Taylor knows that Labor's goal of zero net emissions by 2050 would entail huge subsidies to the renewable sector, and more taxes. It is another way of wedging Labor. Predictably, the Roadmap was criticised by the Clean Energy Council which warned that a technology roadmap that ignored the commercial reality and enormous investment appetite for renewable energy would be 'a waste of time, taxpayer money and a distraction to the energy transition'. Instead, the Roadmap should leverage Australia's comparative advantage around solar and wind energy, as the cheapest sources of electricity generation.

Before considering the real cost of renewable energy, we need to look at other major options floated in the plan. Carbon capture and storage has not, and probably never will be, implemented as a holistic solution in the energy sector. For coal power stations to capture carbon and send it underground in suitable aquifers is cost prohibitive, technically difficult and could burn up to 25 per cent more coal for the same energy outcome. Hydrogen has similar issues. Green hydrogen is prohibitively expensive as it involves separating hydrogen through electrolysis. Huge amounts of energy are required to do this and the return on investment is minimal. So called clean hydrogen, which is produced using coal on the proviso that emissions created in producing it would be captured and stored underground, is also prohibitively expensive and technologically difficult.

The main purpose of the Technology Roadmap is not to canvas these technologies seriously but to push through proposals that make gas the foremost transition fuel. It represents a kind of 'back to the future' moment from the early 2000s. But gas is still too expensive as a base load power fuel with costs at around $90–100 megawatt hours (MWH). Gas could be used in peak stations to provide backup for renewables. This is, however, even more expensive at around $150–170 MWH. These are nonetheless cheaper than the Labor/Greens favourite: battery storage. It remains prohibitively expensive at scale. A report by the Industry Super Fund suggested that it would cost $6.5 trillion to provide battery backup for all of Australia. It is not a viable option and is unlikely ever to become one.

Probably the most important political outcome of the plan was that it obliged the Labor Party to call a truce on the energy wars in favour of a bipartisan approach. Labor offered to join the Coalition in dumping Malcolm Turnbull's National Energy Guarantee and Julia Gillard's Emissions Trading Scheme and even embrace carbon capture and storage, the longtime dream of those arguing for a clean coal future. Notably, both sides of politics baulked at nuclear power. When asked, Federal Labor Leader Anthony Albanese immediately ruled it out while the Coalition essentially hedged: 'maybe, but not now'.

There are many remaining differences between the parties beyond Labor's uncosted 2050 net zero emissions target and the role of renewable energy in achieving it. In the non-energy sector, which is responsible for just under 70 per cent of emissions, the differences between the parties are stark. The Coalition's model is to ramp up the taxpayer funded Emissions Reduction Fund (ERF), rebadged the Climate Solutions Fund, which pays companies to reduce excessive emissions in the agriculture, mining and transport sectors. The Coalition would presumably tweak funding for this model to the extent necessary to meet the emission reduction targets agreed at the 2015 United Nations Paris Climate Conference.

In the lead up to the 2019 election, Labor's Shadow Energy Minister, Mark Butler, flagged abolishing the ERF in favour of forcing companies to pay for emissions reductions themselves (or buy carbon credits) to meet the net zero emissions by 2050 target. This would have involved huge costs on industry without an ERF. For example, BHP had direct CO_2 emissions in 2018–19 of 3.51 million tonnes and Bluescope emitted 6.32 million tonnes. The costs of reducing these to zero, or buying carbon offsets or carbon credits to an equivalent amount, would be progressively greater in the lead-up to zero emissions by 2050. Before the 2019 poll, Butler failed to provide any modelling of these costs to business or to the overall economy in both the energy and non-energy sectors arising from Labor's 2050 target. Failing to provide an adequate explanation left Bill Shorten high and dry when questioned by the media. Without costings, his only defence was to keep saying that inaction was more expensive. The electorate were not persuaded.

It was not clear at the time of writing when Labor abandoned its opposition to the ERF. What is clear, however, is that without full costings and a roadmap to achieve the 2050 target, Labor will once again find itself in politically, if not economically, unsustainable territory. If I were the Federal Labor leader, I would not be relying on Butler to deliver these costings nor to convince the electorate that Labor has a viable way forward. Meanwhile, the one credible way that would allow Australia to move to net zero emissions by 2050—in combination with renewables—is nuclear power. Both parties have baulked at this option.

The real cost of renewables

Neither major political party is prepared to question the cost assumptions that undergird claims made by both the Greens and the renewables sector. Following the announcement to dump the NEG and ETS, Anthony Albanese justified the decision by asserting that renewable energy was now the cheapest form of power. Consequently, he explained, Australia no longer needs an ETS nor a NEG. What he and the Greens do not say is that they are referring to the 'levelised' or production cost of power and not the 'systems' cost. The latter is the cost of providing power to the places where it is needed on a 24/7 basis. The levelised price does not take account of the cost of the excess capacity and storage needed to facilitate 24/7 power availability.

Professor Freidrich Wagner of the Max Planck Institute in Germany published a study in the *European Physical Journal* in 2016 detailing what Germany had to do for 100 per cent of its electricity requirements to be provided by intermittent renewables. It would have to install over-capacity amounting to 89 per cent more than needed. Why? To store sufficient energy in batteries or other storage modes for use when renewable power was not available. There are also significant costs of new transmission lines to bring renewable power to the end user. If the additional system costs of storage and transmission are taken into account, the cost of renewables significantly surpasses the other low-cost base load power sources that are currently available. Despite claiming to be the cheapest energy source, the renewable energy sector continues to call for the retention and expansion of the renewable energy target and its associated subsidies. Further, the sector has successfully acquired subsidised power purchase agreements from state governments. The truth is self-evident: if renewables were really cheaper on a system basis they would not seek subsidies. In dumping the ETS, Labor has flagged its support for expanding two entities, the Clean Energy Finance Corporation and the Australian Renewable Energy Authority, both of which subsidise the renewable sector, and foreshadowed further support for subsidised power purchase agreements outside the market.

Here is one way to look at the challenge: when renewable energy is available in large quantities, meaning the wind is blowing and the sun is shining, there is often an oversupply of power. Renewable energy operators respond to this oversupply by market bidding at a price approaching $0 and even negative dollars. They can do this because if they 'sell' their power to other sources they will still receive a subsidy of about $30 MWH which a coal station, for example, does not receive.

Consequently, Labor and the Coalition will continue to be pressured to extend the existing renewable energy subsidy beyond the current 33,000 GWH and the 2030 cut off. If there is one thing about which the major parties can agree, it is continuing the smokescreen concealing the real cost of renewables. The Coalition has embarked on massive, high-cost projects that are specifically designed to provide backup sources of power when renewables are unable to meet demand. Prime Minister Scott Morrison went to Tasmania two days before the 2019 election promoting his hydro 'battery of the nation'. It probably cost Labor two seats in northern Tasmania and, ultimately, the election. The promise included building a new 1200 megawatt 'Marinus Link' to Victoria. The 'Basslink', which carries about 300 mega watts, was built by the Bracks Labor Government when I was the state's energy minister.

Morrison also promised associated transmission upgrades at a cost of $A3.5 billion and construction of up to 12 hydro pump facilities capable of holding 4800 megawatts of hydro power and, with new renewable energy infrastructure, the overall cost could exceed $A8 billion. Former prime minister Malcolm Turnbull's 'Snowy 2.0' hydro project will cost an estimated $A10 billion. At full capacity it will be able to supply an extra 2000 megawatts but only for seven days. After a week, the water will need to be pumped back up the system using 40 per cent more power than the amount originally generated. The benefits of both 'battery of the nation' and 'Snowy 2.0' projects are plainly exaggerated while the high additional costs of propping up renewable energy is never factored into calculations for the entire cost of renewables.

The case for nuclear

The major political parties also seem to agree about leaving nuclear energy outside the energy debate. I called for a mature debate on nuclear power in the Victorian parliament during 2010. I argued that if climate change was the defining moral challenge of our generation, we could not rule out the only form of base load power that produced zero carbon emissions. A year later, a massive tsunami/earthquake led to the Fukushima nuclear disaster in Japan, setting back any reasoned discussion of nuclear power for years.

Countries like Germany reacted to Fukushima by promising to phase out nuclear power production by 2022. As a result, Germany's coal power stations will have to have their life extended to, at least, 2038. How does it make sense to close non-carbon emitting nuclear power stations and keep polluting coal-fired stations open? Even with a massive $A580 billion investment in renewables since 2009, Germany has only managed to flat line its carbon emissions despite a 50 per cent increase in electricity prices. Germany is able to shut down its coal and nuclear stations because it is geographically located in a way that allows it to import nuclear and coal fired power from its neighbours, principally France, Poland, Belgium and the Nordic states. It is also able to access huge amounts of gas from Russia. With 27 per cent of its electrical power coming from renewable sources, it is certainly no renewable energy 'Nirvana' as some commentators have claimed.

Contrast the German experience with that of France. It produces one-tenth of Germany's emissions and energy consumers are paying half the price. It does this by meeting 72 per cent of its energy needs through nuclear power generation. With the introduction of new technology, France is also recycling 17 per cent of its nuclear waste. This 'waste' is producing more power while the volume of waste is reduced. The French versus German experience runs counter to the claim that nuclear power is too expensive. Overlooking the French experience, a CSIRO study put the cost of power from the new stable of Small Modular Nuclear Reactors (SMRs) being built all over the world at a hefty $A251-$A330/MWH. This is an example of how political ideology has infiltrated even respected organisations like the CSIRO. The CSIRO report has been strongly criticised by the Bright New World (BNW) group because the technology it cites bears no relationship to the SMRs which are closest to commercialisation, assumes extremely high capital costs and accepts unrealistically low capacity factors.

The BNW paper addresses these shortcomings by using more realistic estimates. It concludes that figures of $A85-A$97/MWH are more realistic. The CSIRO report also assumes a discount rate/cost of capital of seven per cent and a build time of five years. Using a more realistic discount rate/capital cost of three per cent (given current low interest rates), BNW estimates the cost could be as low as $A60MWH. A second report from NuScale, a company closest to regulatory approval and commercial deployment of an

SMR, criticised the CSIRO report for not reflecting its costings. NuScale insists they are $US65 MGH. They make the point that no-one from the CSIRO even bothered to contact the company when preparing their report.

The safety of nuclear reactors, especially the SMRs and the Gen III+ reactors, is also worth noting. These reactors now incorporate passive inherent safety features. The European Pressurized Reactor System (PRS), for example, that is incorporated in China's new Taishan 1 has four safety systems, each capable of 100 per cent containment. The Westinghouse AP1000 Gen-III+ reactor has a core cooling system including passive residual heat removal by convection and improved containment. It has been built in China at Sanmen and is under construction at Vogtle in the United States. These units are being assembled from modules which better fit the trend to SMRs.

SMRs are in advanced stages of production around the world and Australia falls further behind with every year of inaction. An arrangement using designs from Rolls Royce in Britain or designs from Nuscale, GE or Westinghouse designs in the United States would put Australia at the forefront. The Rolls Royce model, for example, is a fully enclosed SMR that has multiple 100 per cent passive and active containment systems and earthquake and missile protection. It delivers 450 MGH of continuous power. Rolls Royce claims that it can deliver power at a cost of $A75 MGH. If true, it is cost competitive with coal and certainly with gas. The new designs are also capable of being turned down to as low as 30 per cent output (unlike coal) and are therefore perfect partners for intermittent renewable energy. Nuclear power is also safe. A longitudinal study conducted by *Forbes* in 2018 put the number of deaths per thousand terawatt hours of electricity produced worldwide as coal 100,000, natural gas 4000, Hydro 1400, rooftop solar 440, wind 150 and nuclear 90.

Late in 2019, the Morrison Government established a cross-party parliamentary inquiry to consider the preconditions for a nuclear industry. It resulted in a cautious approach and findings from Coalition members in support of Small Modular Reactors. The minority Labor report opposed the findings and dismissed the potential for Small Modular Reactors despite their being in advanced stages of development in the United States, Canada, Europe and China. Despite the committee's findings in support of continuing considering the establishment of a nuclear industry, the Morrison Government has put further action on hold for a decade and the Labor Party has outright rejected any follow-up activity. A real bi-partisan consensus is simply not possible without nuclear because only nuclear can get Australia anywhere near net zero emissions by 2050. A bipartisan vision could include not only nuclear power stations to achieve this goal but international cooperation in developing technology that used Synroc—a 'synthetic rock' pioneered by an Australian research team in 1978—to produce cylinders of dense, synthetic rock containing nuclear material that can be buried in safe locations. Australia could also exploit ANSTO's existing expertise to enhance safeguards and to participate in the production and commissioning of SMRs over the next 10–15 years. A nuclear Industry would provide jobs, access to new technology and lower carbon emissions.

Nuclear propelled submarines

A bipartisan political approach to nuclear energy would not only deliver a credible net-zero-emissions-by-2050 policy, it might also allow the Government to revisit Australia's most substantial Defence acquisition. I am sceptical about both the future submarine's price tag—which will probably be higher—and the effectiveness of submarines that are conventionally propelled—which will certainly be lower.

The French nuclear version is capable of more than 30 knots submerged, has unlimited range and can operate for a decade without refueling. Similar American nuclear submarines can achieve speeds of 50 knots and need no refueling for 25 years. Our diesel electric submarines will have a top speed of 20 knots; similar to our ageing fleet of *Collins* class submarines. Derek Woolner noted that 'by the time HMAS *Attack* (the first of the new submarines) hits the water in the early 2030s, it's going to be obsolete'. The costs of converting the French design to diesel electric propulsion means our submarines will cost at least $A4.8 billion each. French and American nuclear submarines are estimated to cost half that amount. We could renegotiate the extant contract to allow some of these submarines to be nuclear while retaining a significant local production element. Even allowing for a $A400 million fee to alter the contract, the overall price would still be cheaper.

Hugh White has argued that we need fewer operationally vulnerable and financially expensive surface warships and more state-of-the-art submarines to form an unseen deterrent force in our northern waters. His thinking runs counter to the Government's latest $A270 billion Defence spending proposals which focus principally on warships. The submarine project became a political football between Malcolm Turnbull and Tony Abbott, with Abbott preferring the Japanese version but Turnbull opting for the French design on becoming prime minister. Neither Abbott nor Turnbull would consider nuclear powered submarines and nor would the Labor Opposition at the time. Abbott has since claimed that our new submarines 'will have less power, less range, less speed and less capability than the existing (nuclear) submarine on which it is based and it will come into service about a decade later than would be optimal.' He is right but his conversion to nuclear powered submarines is too late. I personally find it hard to believe that the remnants of a 1980s-1990s anti-nuclear ideology will require Australia to send its sons and daughters into harm's way without the very best technology accompanying the very best training they receive.

The way ahead

Tentative moves toward a bipartisan energy policy to end the so-called energy wars are doomed to fail. Quite apart from the underlying philosophical differences between the two major parties, there is the ever-present practical temptation to exploit these differences to demonstrate poor policy development or inadequate cost estimates. The lack of clarity and the absence of agreement on both the timing and cost of achieving the net zero emissions target will continue to be a source of deep division. A possible way through this malaise is for the major parties to agree that nuclear power be considered a viable way of reaching the zero emissions target as well as providing backup for intermittent renewable energy. In introducing the nuclear option into the energy debate, a re-evaluation of the submarine project could also occur. A nuclear industry involving safe storage, nuclear power stations and nuclear propelled submarines would increase our wealth and create thousands of highly paid jobs in construction and continuing maintenance. Importantly, a nuclear industry would enhance our defence capability and our standing in the world. It would also dramatically reduce Australia's carbon footprint. This may be a very big ask but all it would take to succeed is political leadership and bipartisan agreement.

3 Australia's past, present and potential nuclear capability

Lyndon Edwards

The birth of an industry

Whilst it is well known that large scale Nuclear Science and Engineering was first employed during the Manhattan Project to produce the Atomic Bomb, it less known that the principle of nuclear power production was envisaged from the first discovery of nuclear fission. The first technical description of a practical nuclear weapon was written by Otto Frisch and Rudolf Peierls while they were both working for the eminent Australian physicist, Mark Oliphant, at the University of Birmingham in Britain at the beginning of the Second World War. Oliphant took the Frisch-Peierls memorandum to Sir Henry Tizard, the chairman of the Committee for the Scientific Survey of Air Warfare (CSSAW). A special subcommittee of the CSSAW, known as the MAUD Committee, produced two reports on the 'Use of Uranium for a Bomb' and the 'Use of Uranium as a Source of Power'. During the war, work concentrated on the first possibility.[1]

After the war, concerted efforts were made to develop nuclear reactors as a source of power, both for naval propulsion and for electricity production. Focussing initially on submarine reactors, American developments were first on sodium-cooled reactors and subsequently on light water cooled reactors (LWR), both using enriched uranium. Canada and Britain focused on civilian power reactors. As they did not have access to enriched uranium, they developed reactors using natural uranium that were cooled by pressurised heavy water (CANDU) and carbon dioxide gas (MAGNOX, AGR) respectively. In subsequent years the search for higher thermodynamic efficiency led to research and development including construction of higher temperature prototypes and demonstrator reactors using gas, liquid metal and molten salt cooling but only the three reactor types in the previous paragraph have been deployed on an industrial scale. Indeed, later adoption of LWR technology by Japan, Korea, France and much of the rest of Europe resulted in most of the world's operating nuclear reactors being of this type.[2]

Atomic energy and Australia

In the post-war period nuclear science and technology were developing rapidly. The Menzies Coalition Government was keen to ensure that Australia was able to access many of the benefits. The Australian Atomic Energy Commission (AAEC) was established in November 1952 to exploit the peaceful uses of nuclear energy. It gained a statutory foundation on 15 April 1953 when the *Atomic Energy Act* (1953) was enacted. The AAEC replaced the two committees that had previously advised the government on matters relating to the development of atomic energy in Australia: the Industrial Atomic Energy Policy Committee (CA 343) from January 1949–April 1952 and then the Atomic Energy Policy Committee (CA 332) from April–November 1952.

The three main responsibilities of the AAEC were to:

promote the search for, and mining and treatment of, uranium in Australia with power to buy and sell on behalf of the Australian Government

develop practical uses of atomic energy by carrying out and assisting research, constructing plant and equipment and employing and training staff; and

to collect and distribute information on uranium and atomic energy.[3]

The first AAEC research staff were sent to the British nuclear establishment at Harwell to gain experience in nuclear science. The AAEC's original offices were at Maroubra in Sydney. When the decision was taken to build a large research reactor, a 70-hectare site was acquired at Lucas Heights, then located beyond the Sydney metropolitan area, to be the site of the first Australian reactor HIFAR (High Flux Australian Reactor).[4] The HIFAR reactor was a modified version of an original UK design called DIDO. The United Kingdom Atomic Energy Authority provided the AAEC with plans and specifications. AAEC modified the original design following early operational experience with DIDO.

HIFAR was the second of two variants of the DIDO class reactor that were built. Three in Britain, one in Denmark, one in Germany and HIFAR in Australia. HIFAR achieved criticality on Australia Day, 1958, and Lucas Heights was officially opened by Prime Minister Robert Menzies later that year. HIFAR was designed to provide a high neutron flux over a large volume, with easy access for irradiating both structural materials and radiopharmaceuticals together with neutron beams suitable for neutron diffraction experiments. The design of the fuel elements was modified over the years to enhance irradiation capability and reactor operation and the level of enrichment was reduced to low enriched uranium (LEU) which has less than 20 per cent U235.[5] The construction of HIFAR enabled Australia to participate internationally in the development of peaceful uses of nuclear energy and in the application of nuclear science and technology to industry, health, education and research. It operated for almost 50 years, longer than any DIDO class reactor. It was finally shut down in January 2007.

In 1961, the AAEC commissioned MOATA, a second, smaller 100 kW Argonaut research reactor. Its purpose was to train scientists in reactor control and neutron physics and to accumulate experimental nuclear data on fuel and moderator systems. While initially used as a research tool and training reactor, the scope of MOATA's operation was later extended to include activation analysis and neutron radiography. MOATA also played an important role in aircraft safety. Commercially, it was used for approximately 15 per cent of all procedures world-wide involving radiography to check the structural composition of jet engine turbine blades. The reactor was also an important tool for Australia's uranium mining industry, providing rapid and accurate measurements of ore composition. MOATA was shut down in 1995 and fully decommissioned and completely dismantled in 2011.[6]

During the early 1960s, the AAEC undertook concept design research and development for a Beryllia-moderated pebble-bed gas-cooled high temperature reactor called ABORIGINE producing world leading work on the nuclear properties of Beryllium and Beryllia. However, together with the rest of the world, its focus changed to water-cooled reactors which culminated in the Australian Government announcing its intention to build a 500 MW nuclear power station at Jervis Bay in 1969. When tenders closed the following year there were 14 offers from the United States, Britain, Germany and Canada. All were subject to a condition that the reactor had to be capable of being fuelled from Australian uranium, with the fuel elements being made in Australia. If uranium enrichment was required, tenders were to show how it could be achieved in Australia. The UKAEA assessment team reduced the choice to four tenders: a pressurised water reactor (PWR) from the United States, a German PWR from Germany and a Canadian pressurised heavy-water reactor before settling on a steam generated heavy-water (SGHW) reactor from Britain. Preliminary engineering design was carried out together with seismic evaluation and initial site clearance construction at Jervis Bay before the project was shelved in 1971.[7]

In the 1970s and early 1980s, the AAEC concentrated its nuclear engineering and technology on the front and back ends of the nuclear fuel cycle. Research and development into enrichment technology was undertaken to investigate how to add-value to Australia's uranium ore assets. The AAEC had advanced centrifuge systems and world leading laser enrichment research and several attempts were made to attract inward investment to commercialise its technology.[8]

At the back end of the nuclear fuel cycle there was also significant investment in the pilot scale demonstration of 'Synroc' technology. Synroc, a particular kind of 'Synthetic Rock', was invented by Ted Ringwood of the Australian National University in 1978. It has since diversified but, generally speaking, is an advanced ceramic comprising geochemically stable natural titanate minerals which have immobilised uranium and thorium for billions of years. These can incorporate into their crystal structures nearly all of the elements present in high-level radioactive waste (HLW) and so immobilise them. It was originally intended mainly for immobilisation of liquid HLW arising from the reprocessing of light water reactor fuel. Synroc is now a more versatile and flexible waste-form technology.

From its inception, the AAEC operated large-scale nuclear infrastructure such as particle accelerators and cyclotrons in addition to its research reactors; it also provided access to its facilities to other researchers. The Australian Institute of Nuclear Science and Engineering (AINSE) was established in 1958 to provide universities and other tertiary institutions with a mechanism to access the unique research facilities at Lucas Heights, and to provide a focus for cooperation in nuclear science and engineering. AINSE's 43 member organisations are presently comprised of one industry partner, two research Institutions and 40 universities from Australia and New Zealand.[9]

In sum, from 1953 to 1987 the AAEC designed, procured, built and operated a broad spectrum of advanced nuclear infrastructure using its experience and talented nuclear scientists and engineers in the pursuit of atomic energy and the use of research reactors to benefit society.

From atomic energy to nuclear science and technology

In 1987 the AAEC was replaced under new legislation by a new government organisation, the Australian Nuclear Science and Technology Organisation (ANSTO). ANSTO initially subsumed most of the existing facilities, projects and programs of the AAEC and, in subsequent years, changed its focus from nuclear power production to being the national centre of competence in nuclear science and technology.[10] Consistent with this change of focus, work on centrifuge enrichment was wound down and the world-leading laser enrichment technology was commercialised through SILEX Systems, which is currently the only available third generation enrichment technology operating at a high technology readiness level (TRL).[11]

ANSTO is currently Australia's national nuclear research and development organisation, and the centre of Australian nuclear expertise. Its specialist expertise is applied to radiopharmaceutical production, research into areas of national priority including health, materials engineering and water resource management and helping Australian industries solve complex problems. It also provides expert advice to government on all matters relating to nuclear science, technology and engineering. ANSTO operates landmark national scientific facilities, including OPAL, Australia's only nuclear research reactor, and the Australian Synchrotron, for the benefit of industry, the Australian research community and all Australians.[12]

While the focus of Australia's national nuclear laboratory changed significantly in 1987, the requirement to develop and sustainably operate large nuclear infrastructure did not. Initially, ANSTO continued to operate both the HIFAR and MOATA reactors but, after review, of Australia's future needs the MOATA reactor was shut down and was put into care and maintenance in 1995. In 1997 the Australian government agreed to provide funds for ANSTO to build a replacement research reactor (RRR) at Lucas Heights.

The RRR functional concept design was developed by ANSTO scientists and engineers and tender issued for construction in August 1999. A contract was placed with the state-owned Argentine company, INVAP S.E., and its Australian alliance partners, John Holland Construction and Engineering and Evans Deakin Industries for the design, construction and commissioning of the replacement reactor on 13 July 2000. The INVAP design was evaluated by ANSTO through a Replacement Research Reactor Project Group (RRRP). This group was responsible for all contractual issues, overseeing the construction, participating in pre-commissioning and acting as the focus for the licensing work. Although the reactor core and associated primary equipment were manufactured in Argentina, the bulk of the reactor and its connected safety and containment infrastructure was manufactured in Australia with an end estimated local content of 62 per cent.[13]

Construction of OPAL, the name subsequently chosen for the RRR, started in 2002. First criticality occurred on 12 August 2006 and first operation at full power (20MW) on 3 November 2006. OPAL is a sophisticated modern research reactor with extensive safety features including two independent shutdown systems based on standard power reactor equipment. All safety system sensors are triplicated and there are also many Engineered Safety Features including a diesel and UPS backed standby power system. Put simply from a safety and operational point of view, OPAL has very many common elements to a modern small modular nuclear reactor (SMR). The principal difference is that the primary product of OPAL are neutrons which are to irradiate materials (for example, to make radio-pharmaceuticals) and under-take scientific and engineering analysis using neutron scattering rather than the heat which is the primary product of a power reactor.[14]

The safe and reliable operation of any nuclear reactor requires an organisa-tional structure that is clearly defined and staffed with competent managers and qualified personnel who have the proper awareness of the technical and administrative requirements and have a positive attitude to safety culture. As OPAL is designed and scheduled to operate for 300 full power days each year, ANSTO requires a substantial number of qualified and experienced nuclear operators and engineers. HIFAR operation overlapped the startup of OPAL to ensure continuity of nuclear radio-pharmaceutical production. To ensure this happened, ANSTO underwent a period of significant reactor operations and maintenance human capital development following guidelines issued by the International Atomic Energy Agency located in Vienna. Unusually for a research reactor, the specification for the OPAL reactor included the provision of a reactor simulator. This enables all OPAL operations staff and engineers to experience aspects of reactor operation and control in the same way as used in nuclear power plant.[15]

It is notable that OPAL, from the original Australian pre-concept design, through a fully competitive tender process, the creation of the Australian-Argentinian industrial consortium, and the construction of the reactor, together with the recruitment and training of a qualified and experienced nuclear workforce took less than a decade. This could not have happened without the deep seated nuclear scientific and engineering expertise embedded in ANSTO and its predecessor, the AAEC. OPAL is arguably the premiere multi-purpose nuclear research reactor in the world. There are older, larger reactors that have potential greater capabilities either to irradiate materials or produce neutron beams for scientific research, but OPAL is the only reactor in operation that combines a world-class neutron scattering facility, which includes a cold neutron source, and world-class radio-pharmaceutical production capability. It operates for more than 300 days annually, providing a reliable source of neutrons for research and radio-pharmaceutical production.[16]

Following the successful commissioning of OPAL, ANSTO utilised its nuclear engineering capability through the creation of large-scale nuclear infrastructure: a large nuclear medicine manufacturing facility to enable growth in its Mo-99 radio-pharmaceutical business and development of a world-leading facility to treat the radioactive waste resulting from Mo-99 production. Mo-99 is used to produce Technetium, the most widely used diagnostic imaging radioisotope that is of critical importance for patient healthcare. In Australia alone, each year 550,000 people receive a diagnosis using Mo-99. Global demand for Mo-99 is large and growing, as more countries develop modern medical systems with up to 40 million procedures world-wide anually.[17]

OPAL was the first research reactor in the world to be built to use both low enriched uranium (LEU) fuel and targets for the production of Mo-99. It is also still the youngest reactor in the world that is used to produce Mo-99 on a commercial scale. In response to closure of some of the few reactors around the world that can produce Mo-99 at a commercial scale, the Australian Government funded ANSTO in 2012 to expand its operations and construct a new nuclear medicine manufacturing plant and an associated waste treatment facility that can utilise OPAL's full potential for Mo-99 production. The ANM (ANSTO Nuclear Medicine) plant was constructed and became fully operational in 2019.[18]

ANSTO's nuclear engineering capability is currently engaged in designing, constructing and operating Symo, the world's first Synroc nuclear waste plant. Building on the original Synroc work I have described, research at ANSTO has designed wasteforms for the immobilisation of nuclear wastes with the original hollandite, zirconolite, perovskite and rutile system still being referred to as 'Synroc' phases in the literature. These phases remain highly relevant in nuclear waste immobilisation today. Variations of the original Synroc formulation were developed for a range of challenging nuclear wastes, including excess plutonium waste streams.[19]

ANSTO has championed the use of hot isostatic pressing (HIPing) for wasteform consolidation and the novel application of this technology is a key aspect of the Symo plant presently under construction. A number of successful demonstrations of Synroc technology have been undertaken; the Symo plant will be the first operational industrial scale Synroc production facility. It is designed and being built by ANSTO and is expected to lead to further implementations of this innovative technology once complete.[20]

Establishing a tradition

From the inception of the AAEC which concentrated on nuclear power to the present day, ANSTO, with its focus on world-class nuclear science and radiopharmaceutical production, has maintained a substantial nuclear engineering capability including the ability to procure, build and operate nuclear reactors and associated nuclear plant. In its heyday in 1967–68, AAEC had around 1000 staff, including 400 science and engineering graduates. There has been only relatively small changes in staffing levels over the years and in

2018–19 ANSTO had approximately 1200 staff, including over 200 engineers with nuclear experience.

Throughout the years and in combination with a number of Australian engineering companies, AAEC/ANSTO has designed, procured, built and operated a series of world class nuclear facilities including three small reactors. While Australia's nuclear engineering capability has never had the industrial capacity to support a major large nuclear reactor manufacturing enterprise, it has had the continuing ability to support the design, procurement and operation of small reactors. This capability remains strong and could be employed to support the use of small modular reactors (SMRs) either for power production or naval propulsion.

Endnotes

1   AE Binnie, *From Atomic Energy to Nuclear Science: A History of the Australian Atomic Energy Commission*, PhD dissertation, Macquarie University, 2003: http://hdl.handle.net/1959.14/44919

2   Binnie, *From Atomic Energy to Nuclear Science.*

3   Binnie, *From Atomic Energy to Nuclear Science*; Philip Baxter, *Australian Atomic Energy Commission, The First Ten Years: 1953–1963*; http://apo.ansto.gov.au/dspace/bitstream/10238/2809/2/First_Ten_Years_1963.pdf

4   Binnie, *From Atomic Energy to Nuclear Science*; Baxter, *Australian Atomic Energy Commission, The First Ten Years.*

5   *Nomination of the HIFAR Research Reactor as an Engineers Australia National Engineering Landmark*: https://portal.engineersaustralia.org.au/system/files/engineering-heritage-australia/nomination-title/HIFAR.pdf

6   BH O'Connor, AR Chivas, DW Mather, JDC Studdert, AE Binnoe, *AINSE, An institute for research and training excellence in nuclear science—The First 50 Years*, 2008.

7   Binnie, *From Atomic Energy to Nuclear Science.*

8   Binnie, *From Atomic Energy to Nuclear Science.*

9   O'Connor et al, *AINSE, An institute for research and training excellence in nuclear science.*

10   Binnie, *From Atomic Energy to Nuclear Science*; Baxter, *Australian Atomic Energy Commission, The First Ten Years.*

11   Third Generation Laser-Based Uranium Enrichment Technology, Silex Systems, http://www.silex.com.au; R Snyder, 'A Proliferation Assessment of Third Generation Laser Uranium Enrichment Technology', *Science & Global Security*,

vol. 24, no. 2, pp. 68–91, 2016:
https://doi.org/10.1080/08929882.2016.1184528

12   ANSTO: the Australian Nuclear Science and Technology Organisation (www.
ansto.gov.au)

13   I Smith, 'Progress on ANSTO's Opal Reactor Project and its Future Importance
as the Centrepiece of ANSTO's Facilities', Proceedings of the 15th Pacific Basin
Nuclear Conference,
www.webetc.info/pnc/2006-Proceedings/pdf/0610015pbncfullpaperoraltempl
ate00286.pdf

14   T Irwin, 'Preparing to Operate the Opal Research Reactor', Proceedings of
the 15th Pacific Basin Nuclear Conference, 2006; www.webetc.info/pnc/2006
Proceedings/pdf/0610015final00240.pdf

15   Irwin, 'Preparing to Operate the Opal Research Reactor'

16   J Chakovski and AT Frikken, 'The Journey of Continuous Improvement in the
Reliability and Availability of the OPAL Reactor', Proceedings of the 8th meeting
of the International Group On Research Reactors, Sydney 2017; www.igorr.
com/Documents/2017-SYDNEY/Presentation/Monday/45%20Continuous%20
Improvement%20Availability%20Reliability%20OPAL_Chakovski_1.pdf

17   SM Rathmann, Z Ahmad, S Slikboer, HA Bilton, DP Snider and JF Valliant, 'The
Radiopharmaceutical Chemistry of Technetium-99m' in J Lewis, A Windhorst
and B Zeglis (eds) *Radiopharmaceutical Chemistry*, Springer, Cham. 2019;
https://doi.org/10.1007/978–3-319–98947-1_18; National Academies of Sciences,
Engineering, and Medicine. Molybdenum-99 for Medical Imaging. Washington,
DC: The National Academies Press, 2016 https://doi.org/10.17226/23563; www.
nap.edu/catalog/23563/molybdenum-99-for-medical-imaging

18   National Academies of Sciences, Engineering, and Medicine. Molybdenum-99
for Medical Imaging. Washington, DC: The National Academies Press, 2016
https://doi.org/10.17226/23563; www.nap.edu/catalog/23563/molybdenum-
99-for-medical-imaging; J Senior, 'Reliable & Timely Mo-99 Supply Globally',
in Proceedings of the Topical Mo-99 Meeting, Montreal, 2017; mo99.ne.anl.
gov/2017/pdfs/presentations/S2-P4%20Senior%20Presentation.pdf

19   *The Australian Nuclear Medicine Project,* (www.ansto.gov.au/products-services/
health/services/ansto-nuclear-medicine-project)

20   ER Vance, DT Chavara and DJ Gregg, 'Synroc development—past and present
applications, MRS energy & sustainability'; *A Review Journal*, vol. 4. E. 2017;
*www.cambridge.org/core/journals/mrs-energy-and-sustainability/article/
synroc-developmentpast-and-present-applications/9EDDFC7539E93CE8241B
A8E7F6F30C38;* DJ Gregg, R Farzana, P Dayal, R Holmes and G Triani, 'Synroc
technology—Perspectives and current status (review), *Journal of the American
Ceramic Society*, 2020, doi: 10/11/11/jace 17322; https://ceramics.onlinelibrary.
wiley.com/doi/abs/10.1111/jace.17322

4 Australian uranium mines and mills

Mark Ho

The processing of ores in Australia for production of uranium has a long, but relatively limited, history. Only nine uranium production plants (mills) having been operated since 1954, and only three of these still producing uranium in 2020. By comparison, during the last 65 years, there have been about 140 operating uranium mills in the world. Including several on stand-by, there are now approximately 40 in the world. Although only a limited number of mills have been operated to recover uranium, many uranium deposits have been discovered and numerous feasibility studies prepared for new operations. The history of uranium discoveries, and the political, market and environmental considerations that have affected development are of equal interest to the differences in the treatment plants constructed and operated to produce uranium oxide concentrate.

What follows is a broad overview of uranium mining and processing followed by survey of the key developments in uranium discoveries and mill construction in Australia. The final part covers treatment plants, highlighting the changes in process flowsheets, process conditions, uranium production and tailings management. I will conclude with an update on recent discoveries and potential new operations and their status.

Uranium mining and processing
Uranium is widely distributed in the Earth's crust, being more abundant than antimony, cadmium, bismuth and precious metals. The average uranium concentration in the earth's crust is 3 parts per million (ppm). The United

States Geological Survey has listed about 160 uranium minerals, with another 70 minerals containing minor amounts (or impurities) of uranium (and thorium). The most economically important minerals are: uraninite/pitchblende (an oxide), carnotite (a vanadate), coffinite (a silicate), autunite (a phosphate) and brannerite (Ti complex oxide).

Uranium ores occur in many deposit types. The International Atomic Energy Agency (IAEA) has classified deposits into 15 types, ranging in geological order from magmatic to sedimentary and surficial. From a current production perspective, the most important types are sandstone, proterozoic unconformity and polymetallic hematite breccia complex. From a mining perspective, the deposits can be roughly classified into two categories. The first is sedimentary: sub-horizontal layers, thickness of several tens of centimetres to several tens of metres. The second are vein type: often sub-vertical with the same thickness as sedimentary.

There are three main mining processes used in the uranium industry, namely open-pit mining, underground mining and in-situ leaching (ISL, also known as in-situ recovery or ISR). The choice of open pit or underground mining depends primarily on deposit depth (U/G > 150–200 m). ISL, developed in early 1960s, is now widely used. It is limited to sandstones with high permeability, depths 50–700 m. ISL involves treating ores in underground aquifers by pumping an aqueous solution from the surface through the orebody to dissolve the uranium mineral. The flowsheet for processing uranium-bearing ore obtained from open-pit or underground mining has not changed significantly in the last 60 years. The flowsheet typically involves crushing of ore, grinding crushed ore to liberate/expose the uranium minerals, leaching with sulfuric acid or carbonate solution, solid/liquid separation, solvent extraction (SX) or ion exchange (IX), chemical precipitation, and drying/calcination of the precipitate to produce a uranium oxide concentrate (UOC)). The flowsheet for uranium mining via ISL, once the uranium is dissolved in the solution pumped to the surface, is identical and typically involves IX, chemical precipitation and calcination.

Uranium mines and mills, 1900–1970

Uranium mineralisation was first recorded Australia in 1894 at Carcoar in New South Wales, where torbenite (uranuim phosphate mineral) was found associated with a cobalt ore. Another early discovery reported in 1904 was an occurrence of euxenite at Marble Bar in Western Australia. Prior to 1944, the only known significant occurrences of uranium mineralisation were Radium Hill and Mount Painter, both in South Australia, discovered in 1906 and 1910, respectively. These ores were first worked intermittently to recover their radium[1] content up to 1934, which was predominantly used in the medical industry. About 2,000 tonnes of ore was treated. The mines were finally closed because of the refractory nature of the ores and the discovery of the rich pitchblende (uranium oxide) deposit at Great Bear Lake in Canada. At that time, uranium had minor commercial interest (for ceramic glazes), and only a fraction of the uranium content was recovered as a by-product.

Ore from Radium Hill was upgraded by magnetic separation to produce a concentrate, which amounted to about 30 per cent of the mass of the ore. The concentrate consisted predominantly of ilmenite and contained about 1.4 per cent uranium. Between 1911–15, a metallurgical facility at Hunters Hill in Sydney processed a concentrate from Radium Hill to generate uranium oxide and radium chloride for sale. A parliamentary inquiry concluded that the plant probably processed about 500 tonnes of concentrate containing 7 tonnes of uranium. Uranium recovery was about 85 per cent.

The targeted search for uranium in Australia began in 1944 at the request of the British Government, when uranium was urgently required for military purposes. In 1948, tax free rewards were introduced by the Commonwealth Government for discoveries of uranium ore, the maximum reward being £50,000. This was followed in 1949 by establishment of a uranium buying pool, and in 1952 by the exemption from income tax of profits derived from mining and treatment of uranium ores. These initiatives led to discovery of the first Rum Jungle orebody (White's) in 1949. Contracts for supply of uranium from both Radium Hill and Rum Jungle were entered into with the Combined Development Agency (CDA) in 1952 and production from both operations commenced in 1954. The Commonwealth Government was provided with funds from the CDA, the joint United Kingdom/United States uranium purchasing agency, for the exploitation of Rum Jungle and Radium Hill.

The Rum Jungle and South Alligator area

The Rum Jungle Treatment plant in the Northern Territory was the first to produce yellowcake in Australia in September 1954. Subsequent exploration of the area led to discovery of uranium at White's Extended, Dyson's, Rum Jungle Creek South (RJCS), Mount Fitch (uranium prospect) and small uranium-copper orebody at Mount Burton. Initial production was from White's, Dyson's and Mt Burton, and about 10,000 tonnes of ore from outside mines, was purchased by the Australian Atomic Energy Commission (AAEC, now ANSTO) for treatment at Rum Jungle. By the end of 1958, White's and Dyson's orebodies had been mined out and provided sufficient ore for the CDA contract. The RJCS deposit was mined between 1961–1963. Treatment of stockpiled RJCS ore extended operation past the CDA contract until early 1971, when the plant was closed and sold.

The uranium concentrate produced after the CDA contract (2050 tonnes U_3O_8) was retained by the AAEC at Lucas Heights. The stockpile of uranium oxide was sold in 1993–94 and 1994–95 for electricity production in nuclear power stations in North America. Open cut mining was used for all orebodies. Rum Jungle was in fact a uranium and copper mine and produced 3530 tonnes of U_3O_8 and 20,200 tonnes of copper, mainly as a flotation concentrate. Lead ores as well as small amounts of zinc and nickel were also mined or stockpiled for later processing. The national benefit from Rum Jungle was substantial. In addition to cash profits and a stockpile of uranium oxide, the operation significantly contributed to developing the Northern Territory and provided experience in mining in monsoonal conditions. It also offered useful lessons for subsequent developments in the Alligator Rivers region.

Following discovery of Rum Jungle, and clarification of conditions relating to uranium exploration, a widespread uranium prospecting boom commenced in 1954, which resulted in the discovery of eleven small high grade deposits over a 20 km interval (including El Sherana, Saddle Ridge, Coronation Hill, Palette, El Sherana West, Rockhole and Teagues) in the South Alligator area (SAA), 215 km east-south-east of Darwin. Of the 11 mines in the area, nine were exploited by United Uranium N.L. (UUNL) and two (Rockhole and Teague's) by South Alligator Uranium N.L. (SAUNL). Total production from the area amounted to 835 tons of uranium oxide from about 144,000 tons of ore, mined from the widely separated small ore bodies. Mining practice varied from small open cuts to gloryholes, open stoping and cut and fill stoping.

The Rum Jungle site was abandoned with little remediation being undertaken, apart from a token effort made in 1976. The mining and processing at Rum Jungle had substantial environmental effects, primarily elevated dissolved copper from Acid Metalliferous Drainage (AMD) which polluted the East Branch of the Finniss River (EBFR). As a result, $A18.6 million was spent on remediation between 1983 and 1986 and the program was hailed as best practice at the time. Unfortunately, the works did not completely resolve all the issues and by 2000 the situation was deteriorating. A series of investigations began in 2004 which eventually resulted in the Northern Territory and Commonwealth governments entering into a National Partnership Agreement in 2009 to complete investigative work to inform a rehabilitation plan, deliver site maintenance and continue environmental monitoring. As a result, a comprehensive draft EIS was prepared in early 2020 by the Northern Territory Government for a project entitled 'Rum Jungle Rehabilitation Project'. Construction is estimated at five years, followed by another five years of stabilisation and monitoring. The project's high level objectives are to improve the environmental condition on-site and downstream of site within the EBFR and to enhance on-site environmental conditions to support future land use, including cultural values.

Radium Hill

Radium Hill and Mount Painter were examined for uranium production in the late 1940s but initially abandoned because of their size and complex process requirements. By 1950, however, a treatment process was developed for Radium Hill ore and there were sufficient ore reserves to justify the installation of a treatment plant. In 1952, a contract was signed by the Commonwealth and South Australian governments and the CDA for supply of uranium oxide from Radium Hill over a seven year period. Underground mining and production of a concentrate by heavy medium separation and flotation commenced at Radium Hill in 1954. The concentrate was transported to Port Pirie for treatment in a plant to produce uranium oxide (capacity 160 tons U_3O_8/year), and recover rare earths, which commenced operation in 1955. The physical and chemical processing plants were dismantled in 1962.

Mary Kathleen

The prospecting boom that saw discovery of uranium deposits in the Northern Territory also resulted in the discovery in 1954 of the Mary Kathleen deposit, near Mount Isa in Queensland. After securing a contract with the UKAEA in 1956, a treatment plant and town were constructed, the former adjacent to the single open cut mine, and commenced operation in 1958. The plant was placed on care and maintenance in 1963, when the contract was fulfilled. The plant was restarted in 1976, and operated until 1982 to fulfil commercial contracts. From 1954 to 1971, the five uranium mills in Australia produced more than 9000 of UOC, which was only a small fraction of global production in this period. The decreased demand in uranium for military-related purposes for the United States and Britain caused the world price of uranium to fall, resulting in closure of most of the mines in the early 1960s as contracts expired. This was accompanied by a near-cessation of uranium exploration in Australia between 1961 and 1966.

1970–2020

The late 1960s brought about a major turn in the Australian uranium exploration and mining industry, where a new governmental export policy encouraged increased activity from exploration companies. This policy was initiated because nuclear power had become established technologically and the Commonwealth Government believed that the time was approaching when the state power authorities would need to give serious consideration to introduction of nuclear power. This exploration effort led to discovery by 1972 of significant deposits, including Beverly and Honeymoon near Lake Frome in South Australia, Ranger, Nabarlek, Jabiluka and Koongarra in the Alligator Rivers region of the Northern Territory, and Yeelirrie in Western Australia. In addition, many other uranium deposits were discovered in the period 1966–85, including Angela, Bigryli, Mount Gee, Oban, Kintyre, Lake Way, Lake Maitland, Centipede, Mulga Rock, Manyingee, Ben Lomond, Beverley and Olympic Dam. These exploration efforts contributed significantly to establishing Australia as the most uranium resource rich country in the world. Since this major exploration effort, only a few significant new mines have been discovered. In 1970–72, Mary Kathleen Uranium, Peko/EZ (Ranger) and Queensland Mines (Nabarlek) obtained commercial contracts for the export of 11,000 t U_3O_8 during the period, 1976–86. As developments in the Northern Territory were delayed by the Fox Inquiry, the early delivery commitments of Peko/EZ and Queensland Mines were met by borrowing from the then AAEC yellowcake stockpile. When production commenced, these borrowings were replaced. Mary Kathleen restarted in 1976 to meet these contract requirements.

Fox Inquiry

The new discoveries occurred at a time of increased environmental awareness. In 1974, the Australian Parliament passed the *Environment Protection (Impact of Proposals) Act.* This Act required all mining companies to have plans for closure, decommissioning and rehabilitation approved before commencement of mine development. One of the first projects to fall under the scrutiny of the Act was the proposal by Ranger Uranium Mines to construct a mine and mill at Jabiru in the Northern Territory. Consequently, a government inquiry examined the impacts of any proposed mining in the Alligator Rivers Region, a region of great environmental and cultural significance. The Ranger Uranium Environmental Inquiry, under the direction of Justice Russell Fox, was appointed in July 1975. Its brief was to inquire into the environmental aspects of a proposal by the AAEC in association with Ranger Uranium Mines to develop Northern Territory uranium deposits. (In 1974, a joint venture agreement was set-up between the AAEC, Peko and EZ. The AAEC interest was subsequently bought out in 1980 and Energy Resources of Australia (ERA) was set-up as the operating company).

The Fox Inquiry issued two reports. The first, released in 1976, dealt solely with the grounds of objection to uranium mining in the Alligator Rivers Region. The second report, issued in 1977, considered the Ranger uranium proposal in detail and dealt with other issues, such as Aboriginal land rights, establishment of a national park in the region, and other environmental matters. The Inquiry found that if uranium mining were properly regulated and controlled, its hazards were not sufficient to prevent development of the mines. The Inquiry recommended establishment of a comprehensive system of environmental monitoring and research, overseen by a coordinating committee representing all the agencies involved, and chaired by a Supervising Scientist, a specially created statutory office. It also recommended the granting of Aboriginal title to a substantial part of the region and creation of a national park, which became the Kakadu National Park.

Three Mines Policy

The early 1980s saw adoption of a Labor Government policy that restricted uranium production to just three mines, namely Ranger, Olympic Dam and Nabarlek, with Olympic Dam granted a concession that would allow continued export of uranium as it was produced incidentally with the mining of copper. Provisional approval for marketing from other prospective uranium mines were cancelled. With Narbarlek finishing production in the late 1980s, Australia did not produce uranium from any new mines until, under a Coalition Government, the start-up of Beverley in 2000. This was despite improved prices of uranium and short-lived bursts of uranium exploration activities in the 1990s. The Honeymoon mine produced a small amount of uranium from 2011 to 2013, but is now in care and maintenance owing to the weak uranium market. Australia presently has three operating mines/mills, namely Ranger, Olympic Dam and Beverley (treating the Four Mile deposit). The Australian Labor Party (ALP) policy persisted as a 'no new mines' policy until 2007 when finally abandoned as ineffective. Opposition to uranium mining was then left to state governments to decide, with Labor governments in Western Australia and Queensland imposing a ban on uranium mining in 2002–2006 and 1989–2012 respectively, before being rescinded later by Coalition governments and banned again in Western Australia in 2017.

Ranger

The Ranger mine commenced operating in 1980, treating ore from open pit No. 1, which was mined out over 1980–95, followed by ore from pit No. 3 from 1997 to 2012. Stockpiled ore was processed after 2012 till 2020. Yearly production has ranged from 2094 to 5678 tonnes of U_3O_8/year. In 2008, additional equipment was installed to allow the processing of about 1.6 million tonnes of stockpiled lateritic ore, which because of its high clay content could not be processed without pre-treatment and application of different leach conditions. In 2007–09, a feasibility study into processing 10 million tonnes per year of low grade ore by heap leaching was undertaken. Approval for the project was sought, with start-up planned for 2014. ERA announced that the plan had been shelved in 2011 owing to high capital costs and uncertain stakeholder support.

Under the present Ranger Authority, processing is to cease by January 2021, and rehabilitation is to be completed by 2026. In 1991, ERA purchased the adjacent Jabiluka uranium deposit (contains around 83,000 t U_3O_8). Following a feasibility study in 1993, Commonwealth approval for an underground mine and mill at Jabiluka was granted, conditional upon all tailings being placed underground. Mine development commenced in 1998–99, but ERA subsequently announced in 2005 that the development would not proceed without agreement from the local Mirrar Aboriginal people. ERA is presently implementing a rehabilitation plan for the site which is expected to cost $808 million. Pursuant to the obligation with the Commonwealth and Northern Territory governments and the traditional owners, ERA will return the Ranger project area to an environment similar to the nearby Kakadu National Park.

Nabarlek

Nabarlek was a small high-grade deposit just inside Arnhem Land, and in the East Alligator River region of the Northern Territory. Agreement was reached with the Northern Land Council and other Aboriginal groups to enable development of the deposit. Queensland Mines commenced operations at Nabarlek in 1979. The Nabarlek 1 orebody was mined out in just over four months of the dry season and 600,000 tonnes of average two per cent U_3O_8 grade ore stockpiled for treatment from 1980–89. During the mining operation, ore and below ore grade material were placed on temporary stockpiles while waste rock was used to construct the permanent stockpile pad and water retention ponds. During the milling operation, the tailings produced were deposited in the mined out pit. After the milling operation was completed in 1988, a small heap leach operation continued for another 12 months. When the ore stockpile was exhausted in 1988, the plant was placed in a care-and-maintenance status while further exploration was undertaken in the expectation that another orebody would be found. This was not the case, however, and part of the plant was dismantled, decontaminated and sold in 1994. Nabarlek was the first 'new' uranium mine to go into production, and to be rehabilitated, totally under Environmental Impact assessment (EIA) legislation. Remediation planning from and before start-up facilitated the whole process and outcome.

Olympic Dam

The Olympic Dam copper and uranium mine, discovered in 1975, is located 560 kilometres north of Adelaide. The massive deposit is 350 metres below the surface and is the largest uranium orebody in the world, containing more than two million tonnes U_3O_8. Production commenced in 1988, at rates of 1500 tonnes annually of U_3O_8, and 40,000 tonnes annually of copper produced mainly via smelting a copper sulphide concentrate. Gold and silver are also produced as by-products. Water for the operation is sourced from the Great Artesian Basin.

The copper concentrate is produced by flotation of the ore, with a portion of the uranium in the ore reporting to the copper concentrate. As the concentrate contains uranium, the common procedure of selling concentrate with precious metals has not been viable as the contained uranium and progeny would create both processing and safeguards complications for a smelter operator. For these reasons, a smelter and associated downstream circuits (for example, electrorefining, anode slimes refinery) were installed at Olympic Dam. As the presence of copper and uranium, at a reasonably high concentration, required additional processing requirements, a pilot plant was operated on-site in 1983–1984 to test the metallurgical processes, with the results obtained informing the design of the full-scale plant.

Between 1996–99, a $1.94 billion expansion was undertaken to increase production to 4,600 tonnes annually of U_3O_8 and 200,000 tonnes annually of copper. Following the purchase of Olympic Dam from WMC by BHP in 2005, BHP outlined a massive expansion plan in 2008 which involved open pit mining and expansion of all processing facilities. The capacity of the expanded operation was to be 19,000 tonnes U_3O_8 annually. An EIS was submitted and approved in October 2011. In 2012, the $A28 billion project was, however, placed on hold while less costly alternatives were investigated. In the same year, BHP commenced investigation of a heap leach process to recover uranium from ore, with the construction of large-scale pilot plant in Adelaide. Following heap leaching, the spent ore is crushed/ground and treated by flotation to recover a copper sulphide concentrate for smelting. The seven-year heap leach trial, completed in 2019, confirmed the viability of the technology at Olympic Dam, which has the potential to join the growth options. BHP is currently progressing studies on the Olympic Dam brownfield expansion project, which is seeking to lift production to 240,000–300,000 tonnes of copper per annum, with a corresponding increase in uranium output. Current copper and uranium outputs are 180,000 tonnes annually and 3565 tonnes U_3O_8 annually. Uranium output has varied depending on plant capacity and grades.

Beverly and Four Mile

The Beverley uranium deposit discovered in 1969 is 520 kilometres north of Adelaide. Beverley is owned by Heathgate Resources, a subsidiary of General Atomics. Beverley is Australia's first ISL mine and the highest grade ISL mine in the world. Production commenced in 2000, and production has ranged from 347 to 2002 tonnes U_3O_8 annually, depending on well field development and age, and ore grades. The deposit is at a depth of 100–130 metres.

The main Beverley deposit consisted of three mineralised zones, north, central and south. As the main orebodies were depleted, the Beverley North project was brought into production in 2009. In 2010 and 2011, recovery from the Pepegoona and Pannikin deposits, respectively, commenced via two satellite plants. At the satellite plants, well field solution is treated by ion exchange to capture uranium, with the loaded resin then trucked to the main processing plant at Beverley. Eluted resin is returned to the satellite plants. Production from all the Beverley deposits virtually ceased during the period 2014–16.

The Four Mile deposit comprises three alluvial fan deposits 5–10 km northwest of the Beverley Mine. Average depth is 180 metres. Four Mile was discovered in 2005 by Quasar Resources, who is affiliated with Heathgate, and it is the best uranium deposit discovered in Australia since Kintyre (Western Australia) in 1985. Following settlement of a dispute between Quasar and a JV partner, production from Four Mile commenced in 2014. The three deposits (East, West and Northeast) contain up to 55,000 tonnes U_3O_8 at an average grade of 0.30–0.34 per cent U_3O_8. Production has mainly been from the East deposit, with production from the West commencing in 2019. Uranium recovery is through the Pannikin satellite ion exchange plant, with loaded resin trucked to the Beverley plant, but work is in progress to install a pipeline to transport pregnant liquor directly to the Beverley plant.

Waste liquors, mainly wellfield bleed solution, are combined, evaporated in lined ponds, with excess liquor injected into selected locations within the mined aquifer. Any solid wastes (pipe/screen scales, contaminated soils) are disposed of in a small repository for low-level radioactive wastes, which has a compacted clay base, lined with HDPE.

Honeymoon

The Honeymoon deposit, discovered in 1972, is about 80 kilometres north-west of Broken Hill. Approvals to produce 450 tonnes U_3O_8 annually were obtained in 1981. Field tests of the acid ISL process were carried out, and a 110 tonnes U_3O_8 annually pilot plant was built on site, but the project was stopped in 1983 as a consequence of changes in Federal government policy in regard to uranium mining after the Hawke Labor Government took office. The demonstration plant and infrastructure were refurbished and on-site tests run between 1998 and 1999, producing about 30 tonnes U_3O_8. The project, which includes other nearby deposits at Gould's Dam and Jason's, was acquired by Uranium One in 2005. Following a new feasibility study, a decision to proceed with the project as a 400 tonnes U_3O_8 annually mine was taken. The projected mine life was 6–7 years at an average grade of 0.24 per cent U_3O_8. Honeymoon commenced ISL operation in 2011. In November, 2013 the operation was placed on care and maintenance by Uranium One, due to the low uranium price and technical issues, which included low uranium leach tenors and problems associated with operation of a novel SX process.

The groundwater at Honeymoon contains up to 10 grams per litre of chloride, higher than at Beverley and other ISL operations. As uranium loading by ion exchange is significantly reduced at these high chloride concentrations, this necessitated development of a custom solvent extraction process using a mixture of anionic and cationic extractants. This process was tested in the pilot plant and implemented in the full scale plant. Honeymoon was purchased by Boss Resources in 2015. A feasibility study completed in 2020 outlined a base case production of 907 t U_3O_8 per year over a 12-year mine life, with a new ion exchange circuit using a new chloride-tolerant resin, and other process improvements, planned to increase capacity. The project can now be fast-tracked to re-start production when the uranium price improves. The feasibility study was based on a Honeymoon global resource of 32,500 tonnes U_3O_8 at an average grade of 0.062 per cent U_3O_8, including Gould's Dam and Jason's.

Production and processing—past operations

Production and process details past uranium mills are shown in Tables 1 and 2.

Table 4.1: Early Australian uranium production#

Operation	Period	Mining	Ore Treated (t)	Average Grade (% U_3O_8)	Production (t U_3O_8)
Radium Hill	1954–62	U/G	970,000	0.7–0.9[1]	850
Rum Jungle (N.T.)	1954–71	Open cut	863,000[2]	0.28–0.4[3]	3,530
United Uranium (UUNL) N.T.	1959–64	O/C and U/G	128,000[4]	0.35–0.68[4]	5204
South Alligator Uranium (SAUNL) N.T.	1959–62	U/G	13,500	1.12	128
Mary Kathleen (MKU) Qld	1958–63	Open cut	2,900,000	0.23[5]	4,080
Mary Kathleen (MKU) Qld—phase 2	1976–82	Open cut	6,300,000	0.1	4,800
Nabarlek N.T.	1979–88	Open cut	600,000	1.9	10,860

#Most data from (Warner 1976)
1 Grade of heavy media/float concentrate, ore was 0.11–0.15 per cent U3O8
2 Not including 275,000 t ore from White's
3 Average grade of White's, Dyson's and Rum Jungle Creek deposits
4 Data for Moline treatment plant, not including pitchblende concentrate
5 Grade after radiometric sorting, ROM grade was 0.15 per cent U3O8

Table 4.2: Uranium processing details

Mill	Port Pirie	Rum Jungle	UUNL* Moline	MKU phase 1	MKU phase 2	Narbalek
U minerals	davidite	uraninite§§	uraninite	uraninite	uraninite	uraninite§§ coffinite brannerite
Grind size	P_{66} 75 µm	P_{55} 75 µm	P_{50} 75 µm	P_{65} 75 µm	P_{50} 75 µm	P_{50} 75 µm
Leach pH	20–50 g/L**	1.1–1.6	0.5–1.0	2.5–1.8	2.0	1.5
Leach time (h)	10	8	15	8	5	24
Acid (kg/t)	350–400	80–90	45	60	60	40
Temp. (°C)	~100		40–45	30–37	55	40–45
Oxidant (kg/t)	none	14	1.5	5	15	2##

Mill	Port Pirie	Rum Jungle	UUNL* Moline	MKU phase 1	MKU phase 2	Narbalek
S/L separation	4 stage CCD	Filtration	4-stage CCD	5-stage CCD	5-stage cyclones and 5 stage CCD	8 stage CCD
U recovery	IX Elute with H_2SO_4	IX, elute with $NaCl/H_2SO_4$ 1962 converted to 4-stage SX extract, with 4-stage NaCl strip	4-stage SX extraction 3-stage strip NaCl Scrub with Na_2CO_3	IX Elute§ with NaCl/ H_2SO_4	4-stage SX extraction 4-stage $(NH_4)_2$ SO_4 strip	4-stage SX extraction 4-stage $(NH_4)_2$ SO_4 strip
U Precipitant	MgO, after Fe removal by lime addition	MgO	MgO	MgO	NH3	NH3
U recovery (per cent)	around 92	90	93	around 90	85	97.5
Tailings treatment	None	None	None	None	None	Neutralised to pH 8
Tailings disposal	Slurry pumped to tailings dams	Slurry pumped to tailings dam, and then to Dysons Open Cut from 1961–65	Slurry pumped to tailings dams	Deposited in dam by beaching, decant liquor to evap. pond	Deposited in dam by beaching, decant liquor to evap. pond	Deposited in mined out pit

\# Radium Hill ore was railed to Port Pirie

* The SAUNL South Alligator mill was very similar to the Moline mill

§§ Phosphate minerals such as torbenite, saleeite, autunite and sklodowskite were in weathered zones

§ A first stage elution was dilute sulphuric acid was used to selectively remove rare earths

** After 10 h at boiling

\#\# Expressed as 100 per cent H2O2, also see Lucas, et al (1984)

All the mills listed in Table 4.2 used sulfuric acid leaching, typically with addition of pyrolusite as oxidant. Apart from Port Pirie, leach circuits were not heated. In fact acid dilution was used at MKU during the phase 1 operation to maintain a leach temperature of 30–37 °C, because of the very reactive nature of the gangue minerals, fluorapatite, allanite and stillwellite (Hartley 1968). In the phase 2 operation, this was discontinued. Where SX was used, alamine-336 was the extractant. Whereas MKU and Nabarlek treated ore from a single pit, the Moline mill treated ore from nine deposits, similarly, the Rum Jungle mill treated ore from several orebodies, and some smaller deposits, as described previously.

As shown in Table 4.2, the processes used at all operations were very similar. There were, nonetheless, some significant differences:

> the refractory davidite mineral in the Radium Hill deposit required severe leaching conditions, that would most likely not be economical in the current market;

> there was a change from ion exchange to solvent extraction when the later technique was established; this switch was also undertaken in South Africa and the United States;

> ammonia precipitation was adopted instead of MgO in later operations;

> the use of radiometric sorting at MKU;

> the use of Caro's acid, a derivative of hydrogen peroxide, as the oxidant at Nabarlek; and

> the methods of tailings treatment and deposition were unusual.

Apart from Radium Hill ore, which contained the refractory U-Ti mineral davidite, the other ores contained uranium, predominantly as uraninite, which is readily soluble in dilute sulphuric acid. Secondary phosphate minerals, which typically require a lower pH than uraninite for dissolution, were also present in some deposits. The Rum Jungle, South Alligator River and Nabarlek deposits were of the unconformity-related type, whereas Mary Kathleen and Radium Hill were metamorphic and intrusive type deposits, respectively.

Production and processing: current operations
While all of the past operations, apart from Port Pirie, were relatively similar, current operations differ in many aspects. The Ranger orebodies are unconformity type deposits located in the Alligator Rivers region of the Northern Territory. The Olympic Dam copper/uranium and the Beverley and Honeymoon in-situ uranium operations are located in the semi-arid inland of South Australia, with all sites only receiving around 190 mm of rainfall per annum, with evaporation of 3000 milllimetres annually. Olympic Dam is a hematite breccia complex deposit, while Beverley and Honeymoon are typical sandstone deposits, like many deposits presently processed by ISL in Kazakhstan.

Production and processing details for Australia's four operating mills are presented in Tables 4.3 and 4.4.

Table 4.3: Uranium production from current mines#

Operation	Start-up	Mining	Average Grade* (per cent U₃O₈)	Total
Ranger N.T.	1980	Open cut	0.27	> 130,000
Olympic Dam S.A	1988	U/G	0.05	87,000
Beverley S.A.	2000	ISL	0.21	8,900
Four Mile	2014	ISL	0.30–0.34	8,205
Honeymoon S.A.	2011	ISL	0.13	312

Most data from WNA (Jan 2020) and IAEA Red Book 2018
From start-up to June 2019
* Notional grade of deposit. Grade will/has varied depending on areas being mined.
** Includes production from Beverley North, Pepegoona and Pannikin.
§ Currently on care and maintenance

The Ranger and Olympic Dam uranium process flowsheets are both conventional, involving acid leach, CCD, SX and uranium precipitation with ammonia. The Olympic Dam flowsheet is complicated by the presence of copper in the uranium leach liquor, which is recovered from the liquor by copper SX, prior to uranium SX. Ranger follows past Australian and South African practice with the use of pyrolusite as the oxidant in leaching, whereas Olympic Dam uses sodium chlorate.

Table 4.4: Uranium process details

Mill	Ranger	Olympic Dam	Beverley	Honeymoon
U minerals	uraninite#, coffinite, brannerite	uraninite, coffinite, brannerite	coffinite	Uraninite, coffinite and U phosphates
Grind size	P_{80} 250 µm	P_{80} 75 µm	n/a	n/a
Leach pH	1.9–2.0	1.3–1.5	1.5–2.0	2–2.3
Leach time (h)	18–24	6–8	n/a	n/a
Acid (kg/t)	30–40	40–60	n/a	n/a
Temp. (°C)	35–45	60–70	30	30

Mill	Ranger	Olympic Dam	Beverley	Honeymoon
Oxidant (kg/t)	5	1	§	§§
S/L separation	7 stage CCD	6 stage CCD	n/a	n/a
U recovery	4-stage SX extraction 4-stage $(NH4)_2$ SO_4 strip	SX extraction in pulse columns 4-stage $(NH4)_2$ SO_4 strip	IX, elute with NaCl/ H_2SO_4	SX, strip with Na_2CO_3
U Precipitant	NH_3	NH_3	H_2O_2	H_2O_2
U recovery (per cent)	around 90	around 75	around 70*	
Tailings treatment	Neutralised to pH 4	none	n/a	n/a
Tailings disposal	Initially to tailings dam, then sub-aqueous to mined-out pits	Sub-aerial deposition in tailings dam, decant liquor to evap. pond	n/a	n/a

\# Primary ore. Weathered and lateritic ore contain saleeite, torbenite and sklodowskite, Collier, et al (2000).

* Ranges from 50 to 90 per cent. Hard to assess, as it is difficult to define the head uranium under leach. Also ore remaining after initial leaching can be releached at a later time.

§ About 80 mg/L of hydrogen peroxide is added to the lixiviant before injection underground.

§§ Sodium chlorate used as oxidant

At Olympic Dam, about 10–15 per cent of the uranium in the ore reports to the copper concentrate produced by flotation. This uranium is recovered in a separate sulphuric acid leaching circuit, before the concentrate is processed in a smelter. The solubilised uranium is subsequently combined with the main uranium leach circuit liquor, from leaching the flotation tailings.

The Beverley and Honeymoon deposits are mined using the in-situ leach process. ISL is the most cost-effective process for the treatment of smaller, low-grade deposits of suitable mineralogy and geological environment. The lack of disturbance of the land and no tailings generation are significant environmental advantages for ISL. Environmental aspects of ISL in South Australia were assessed by Taylor and his colleagues who concluded that Beverley 'has initiated and implemented world best practice methods' and that ISL was 'more cost effective and environmentally responsible than any suggested alternative techniques' (for the Beverley deposit).

In the context of future operations, there are several significant features of operations at the Ranger, Olympic Dam (OD), Beverley and Honeymoon mines.

- Leach liquors at Beverley contain up to 5.5 g Cl/L, which is sufficient to decrease the loading of uranium in IX significantly. OD liquors contain about three grams per litre Cl, which is sufficient to impact on uranium loading in SX. Honeymoon liquors contain 8–10 grams per litre, which necessitated the use of a novel solvent extraction process. A new resin is now available which can tolerate higher chloride concentrations, and offers other operational advantages compared to the solvent extraction process.

- Ranger and ODO produce ADU, which must be calcined at 850° celsius to produce a uranium oxide product. At Beverley and Honeymoon, a saleable uranium peroxide product is produced by drying at 250° celsius, with installation of a calciner now planned for Beverley.

- As Ranger is an open cut operation, low grade and lateritic ore were stockpiled for possible processing towards the end of mine life. The lateritic ore required different processing conditions, compared to typical primary ore. At Nabarlek low grade near-surface material was stockpiled and, after agglomeration, was treated in a small heap leach operation in the last few months before shutdown.

- All tailings (neutralised to pH around 4) at Ranger will eventually be deposited below ground level, while Olympic Dam deposit the fines fraction of the acid tailings in above ground dams, which currently occupy about 400 hectares. Acidic liquor decanted from the settled tailings is recycled to the process. The coarse faction of the tailings and waste rock are mixed with cement and used as backfill for the mine.

- The grind size at Olympic Dam is quite fine, by typical uranium ore practice, as the ore must be ground to a size to liberate the copper minerals in flotation. About 85–90 per cent of the uranium in the ore reports to the flotation tailings, which are treated under the leach conditions described in Table 4.4.

- Uranium recovery at Olympic Dam is relatively low because of the higher proportion of non-uraninite minerals and the complex leach solution chemistry.

- At Beverley, uranium is directly precipitated using hydrogen peroxide from IX eluate. A first stage precipitation step is carried out to remove iron. At Honeymoon, the carbonate SX strip liquor is first acidified to destroy the carbonate, with uranium then precipitated with hydrogen peroxide.

The Jabiluka and Ranger 3 Deeps are attractive projects for potential future development.

A number of uranium projects received State and Federal approval in the period 2015–2019. All of these deposits are in Western Australia. All projects are on hold awaiting an increase in uranium prices. Project details are set out in Table 4.5. While the Kintyre deposit will be processed by a conventional acid leach, solvent extraction route, the other three projects warrant additional comment.

Table 4.5: Approved potential uranium projects#

Project	When discovered	Contained Uranium (tonnes U_3O_8)	Average grade* (per cent U_3O_8)	Likely production (tonnes U_3O_8/year)
Jabiluka	1971–73	> 135,000	0.55	???
Ranger 3 Deeps	2008	34,000	???	???
Kintyre	1985	25,270	0.62	3,180
Wiluna	1972–77	38,000	0.048	900
Mulga Rock	1979	40,000	0.057	1,560
Yeelirrie	1972	57,760	0.16	3,890

\# Data from WNA 2020 and Red Book 2018
* May not represent the feed grade of ore to the processing plant

Yeelirrie and Wiluna (includes Lake Way, Centipede-Millipede and Lake Maitland) are both near surface, calcrete-hosted uranium deposits, containing carnotite, with high clay levels, and with saline local water. Owing to the high carbonate content of the ores, atmospheric alkaline leaching was chosen. As the high chloride content of local water precludes use of IX, uranium will be precipitated directly from the carbonate leach solution as sodium diuranate (SDU). The SDU is then dissolved and uranium re-precipitated as uranium peroxide. As ores of this type have not previously been treated commercially for recovery of uranium, the Yeelirrie flowsheet was tested in a fully integrated pilot plant that was operated for over 12 months in 1980–81 in Kalgoorlie.

Mulga Rock is a carbonaceous (lignite) sedimentary hosted deposit that can be readily upgraded by gravity methods to reject a uranium depleted silica rich sand fraction. As the uranium is subject to preg-robbing by the lignite, the ore will be processed by an acid leach/RIP process, where the resin counters the adsorption by the lignite. Vimy groundwater is high in chloride (9–10 grams per litre), requiring the use of a more chloride tolerant weak base resin. Uranium is eluted from the resin using sodium chloride which is then treated by UF/NF to recover sodium chloride for recycle, and produce an upgraded pregnant liquor as feed to uranium peroxide precipitation.

Other projects which have seen some development work over the last 10 years are listed in Table 4.6. Most of these projects would most likely require extra tonnes of uranium to make them viable, except those amenable to ISL recovery.

Table 4.6: Possible future development #

Deposit	Location	When discovered	Contained Uranium (approx. tonnes U_3O_8)	Average grade* (per cent U_3O_8)
Valhalla, Skal, Anderson's Lode	Qld	1954	43,000	0.08
Westmoreland	Qld	1956/59	23,000	0.085
Angularli	N.T.	2012	11,750	1.29
Angela	N.T.	1973	11,750	0.13
Bigryli	N.T.	1973	12,200	0.082
Samphire*	S.A.	2007	21,000	0.023
Manyingee*	W.A.	1974	12,000	0.08
Carley Bore*	W.A	2010	7,100	0.031
Yanrey*	W.A	2010	7,100	0.028

Most data from WNA 2020 and Red Book 2018

* May not represent the feed grade of ore to a processing plant.
* Potential for ISL processing

Conclusion

Uranium was first discovered in Australia more than 125 years ago with continuous mining and milling over six decades. Australia has vast natural resources with South Australia alone possessing 25 per cent of the world's known economically recoverable uranium. With such plentiful and accessible raw material, there is considerable potential to value-add through enrichment and production of nuclear fuel. As all of the uranium currently mined in Australia is extracted for export, a domestic nuclear power industry and nuclear propulsion for Australian submarines might represent a business case tipping point to stimulate a new domestic industry involving all stages of the nuclear fuel cycle. The potential is limitless and the possibilities endless. Leadership is the only absent commodity.

Endnotes

1   Ore containing 1 tonne of uranium contains only 0.3 g of ^{226}Ra.

Part Two

5 Nuclear fuel cycle: International and Australian practices

Tony Irwin

There are over 800 operating nuclear reactors in the world. Some 439 are used for generating electricity in civil power stations, 220 are research reactors and the majority of the remainder are used for propelling 160 ships powered by 200 small reactors, notably some ice breakers, aircraft carriers and, the majority, submarines. Regardless of their application, all nuclear reactors depend on fuel produced by the nuclear fuel cycle. The nuclear fuel cycle involves several processes, however these can be divided into two groups, one for the front end (fuel for the reactor) and the other being the back end of the cycle (management of used fuel). Each group presents challenges and various countries have adopted different approaches or technologies relevant to each phase. Australia, with the world's largest economically recoverable reserves of uranium, vast open spaces, advanced economy and high education standards, has capability in some phases and opportunities in some others. This chapter describes each of the processes of the nuclear fuel cycle, identifying the different options and opportunities.

The 'front end'
The 'front end' includes all the activities from mining uranium, through conversion, enrichment and fuel fabrication up to the point where the fuel is ready to be loaded into a nuclear reactor.

Uranium as mined contains 99.3 per cent uranium-238 and 0.7 per cent uranium-235. Only the uranium-235 is fissionable (that is, it can be used as a fuel) in most nuclear reactors. The majority (around 90 per cent) of power reactors require the proportion of U-235 to be increased (enriched) to 3–5 per cent. Reactors using heavy water (Canadian CANDU pressurised heavy water reactors) or graphite (early British Magnox reactors) as moderator can use natural uranium as the fuel without enrichment. Research reactors typically use fuel enriched to nearly 20 per cent because they want as many neutrons as possible. Research reactors are 'neutron factories' producing neutrons for scientific research, materials research to support civil or military programs and irradiation of materials to produce medical or industrial radioisotopes.

Twenty per cent enrichment is the internationally recognised boundary between Low Enriched Uranium (LEU) and High Enriched Uranium (HEU). Nuclear weapons were assembled with uranium enriched to up to 90 per cent. Nuclear powered submarines have historically used HEU, to enable a long period before refuelling is required. The latest French nuclear powered submarines use LEU, but the core has to unloaded every 8–10 years for reactor pressure vessel inspection, so this fits in with the shorter life of an LEU core. Enrichment requires the uranium to be first converted into a gaseous form to allow physical separation of the uranium isotopes U-235 and U-238; this is the *conversion process*. There is less than 1 per cent difference in mass, which makes physical separation difficult but physical separation is normally used because isotopes cannot be easily separated by a chemical process.

Conversion

In a conversion plant, the uranium is converted into uranium hexafluoride UF_6. Uranium hexafluoride was chosen because fluorine has only one isotope and UF_6 is a gas at reasonable temperatures and pressures (triple point $64°$ celsius, 1.6 bar). The disadvantages of UF_6 are that it is highly corrosive, toxic and reacts violently with water and organic materials.

Conversion is a simple chemical process. As the South Australia Nuclear Fuel Cycle Royal Commission found, 'the most significant safety and environ-mental risks are posed by toxic, corrosive and potentially explosive chemicals rather than the radioactivity of the materials.'[1] The conversion steps are: refine the uranium ore concentrate (UOC) from the mine, particularly to remove any neutron absorbers—dry and reduce to UO_3—reduce with hydrogen to UO_2. This product now can be used in a reactor fuelled by natural uranium.

Further steps are necessary to prepare for enrichment: convert to UF_4 with hydrofluoric acid, then add fluorine gas to convert to the final product UF_6.

The conversion market is dominated by four major players—Cameco (Port Hope Canada); Coverdyn (Metropolis, United States); Orano (Malvesi and Tricastin France) and TVEL (Russia). China is moving towards self-sufficiency in the nuclear fuel cycle and has a plant at Langzhou. While Australia could technically produce UF_6, there is over-capacity in the market and conversion is the least expensive part of the front end processes. The conversion process ends with liquid UF_6 drained into IP2 (industrial package) cylinders. The UF_6 is allowed to cool and solidify for transport. Around 9,000 cylinders, each holding 12.5t UF_6 are transported from conversion plants to enrichment plants every year.

Enrichment

The enrichment process cannot 'create' U-235. It increases the proportion of U-235 in the product and consequentially reduces the proportion of U-235 in the tails (waste stream). Enrichment services are sold in kilograms of SWU (Separative Work Units),[2] a formula which represents the work done to change between different enrichments.

For many years the main technology was *Gaseous Diffusion*. UF_6 gas is forced under pressure through a membrane and, in accordance with Graham's Law, the rate of diffusion of a gas is inversely proportional to the square root of its molecular weight, so that more of the lighter U-235 than U-238 passes through the membrane. The theoretical separation factor is 1.0043, however, meaning that up to 1400 stages are required just to get to 4 per cent enrichment for a power reactor and the process is very energy intensive, typically 2500 kWh/SWU. The last major gaseous diffusion plant in the Western world—Paducah in the United States—shutdown in May 2013. Paducah had 1760 stages to enrich to 2.75 per cent, maximum power demand of 3040 MW and the main process building was 335 metres long, 295 metres wide and 27 metres high.

Centrifuge technology

Centrifuge technology, with a power requirement of less than 50kWh/SWU and a separation factor of greater than 1.3 has replaced gaseous diffusion for enrichment. The world leader is URENCO, a company established on 4 March 1970 by the governments of Netherlands, Germany and the United Kingdom.[3] URENCO has enrichment facilities at Capenhurst (United Kingdom), Gronau (Germany), Almelo (Netherlands) and New Mexico (United States). Tenex in Russia has major enrichment capacity and China is rapidly increasing its capability. Centrifuge technology is tightly held. If Australia decided to establish an enrichment facility it would likely build a URENCO plant under licence. When Orano (formely AREVA) in France decided to replace their giant Tricastin gaseous diffusion plant, it chose URENCO centrifuge technology. There is currently over-capacity in the enrichment market and it is not expected that any more major plants will be constructed. Enriched UF_6 is transported in Type B cylinders holding 2.27 tonnes of UF_6. The cylinders are surrounded for transport with a shock absorber to prevent damage in an accident.

High-essay low enriched uranium

Commercial enrichment plants are normally licensed to enrich to a maximum of 5 per cent. Uranium enriched to greater than 5 per cent and less than 20 per cent is known as High-Essay Low Enriched Uranium (HALEU). The simplest method of producing HALEU is by down blending ex-weapons program HEU. There is a growing need for HALEU for:

> nuclear power plants requiring fuel enriched to greater than 5 per cent for longer burn-up (longer period between refuelling);
>
> reactors operating on LEU for nuclear propulsion;
>
> research reactors designed for 19.75 per cent enriched fuel (for example, OPAL); and
>
> fuel for Gen IV designs, for example TRISO fuel for the Very High Temperature Gas Reactor (VHTR)

Laser enrichment

Research into enrichment using lasers was carried out in the United States, France and Japan in the 1970s. It was abandoned because it was not commercially viable. In the molecular process, the laser has to be tuned to selectively excite U-235, but not U-238 in UF_6. A second laser dissociates the excited $^{235}UF_6$ to $^{235}UF_5$ which precipitates out and can be collected.

In the 1990s, Silex Systems, an Australian company spun off from ANSTO, developed a laser technology with commercial promise. Since this could not be developed in Australia due to the law prohibiting the construction of an enrichment plant, a commercialisation agreement was signed with GE-United States in 2006. Subsequently, in 2008 Cameco bought into the GLE joint technology project and in 2009 a test loop was operating successfully at Wilmington, North Carolina. After 2011, when the enrichment market collapsed, GE-Hitachi announced its intention to exit GLE. Silex and Cameco are trying to keep the project alive awaiting the cost of enrichment returning to pre-2011 levels. Potentially, the laser enrichment technology has an exciting future.

Fuel fabrication

Enriched UF_6 is a universal product used in any reactor. The next step in the nuclear fuel cycle is to manufacture a fuel assembly containing enriched uranium for a particular reactor. A fuel assembly (FA) has four components:

fissile material (U-235, U-233, Pu-239);

host matrix (U-238, oxide/metal);

fuel cladding to prevent the release of fission products; and

structure to enable the fuel assembly to be loaded and unloaded from the reactor

A FA will typically be in a power reactor for up to six years. It has to withstand high radiation and temperature in normal operation and accident conditions. The design has to consider requirements for manufacture, transport, storage, refuelling and reprocessing.

Fuel for a power reactor is typically UO_2 ceramic oxide with a melting point of 2870° celsius, in the form of a fuel pellet around 10 millimetres diameter and around 12 millimetres long. The pellets are contained in a Zircaloy can. Fuel cans are assembled into:

fuel bundles up to one metre long loaded loose into a horizontal CANDU reactor;

fuel bundles joined by a tie bar and loaded vertically into an AGR or RBMK; and

17 x 17 array, 4 metre high for a pressured water reactor (PWR).

When cylinders containing solid enriched UF6 are delivered to a fuel fabrication plant, they are first loaded into an autoclave to change the UF6 into a gas which is then converted in a rotary kiln to UO2 by heating with hydrogen and water. The UO2 powder is compressed into pellets, sintered at 1750o celsius for up to five hours, ground to size and loaded into Zircaloy cans. The cans are filled with helium and the ends welded. There are around 310 fuel pellets in a typical 3.8 metre long PWR fuel can. A typical PWR fuel assembly is around four metres high with a square 17 x 17 array. The structure consists of top and bottom nozzles connected by 24 guide tubes for the control rods. The 264 fuel tubes are located in the top and bottom nozzles and supported by several grids spaced along the tubes.

The fuel fabrication market is also dominated by several big firms. Framatome has plants, particularly for PWR fuel, in France, Germany, Belgium and the United States. GNF (Global Nuclear Fuels) GE with Hitachi/Toshiba specialises in BWR fuel and has plants in the United States and Japan. Westinghouse plants are located in the United States, the United Kingdom and Sweden. Russia and China also manufacture fuel. Historically, fuel assemblies were supplied by the reactor vendor, but the reload market is now very competitive and fuel manufacturers are offering a variety of designs. There is currently over-capacity in the market. It is unlikely that Australia would independently establish a fuel manufacturing facility, but one of the existing manufacturers could set up a plant in Australia.

Accident tolerant fuel

Since Fukushima, fuel that has better performance in accident conditions is being developed, particularly under the United States Department of Energy's *Enhanced Accident Tolerant Fuel (EATF)* program. There are developments in fuel and cladding to give better mechanical strength at high temperatures, more resistance to radiation and corrosion, and less potential for hydrogen production in an accident. For example, Areva is developing chromia doped fuel pellets and chromia coated fuel cladding. Lead test assemblies were loaded into a commercial reactor in 2019. Westinghouse is developing silicide fuel. Accident Tolerant Fuel would obviously be of interest for nuclear propulsion applications.

Tri-structural isotropic fuel

A completely different form of fuel was developed in the 1970s for Very High Temperature Gas Reactors (VHTR) and is now being manufactured for the Gen IV reactor program which includes the VHTR. The fuel is in the form of coated microspheres of LEU. A fuel kernel with a diameter of around 0.5 millimetre is coated with 3 layers of pyrocarbon and one layer of silicon carbide to form a 1 millimetre diameter fuel particle. The fuel particles are assembled into cylindrical rodlets, 12.5 millimetre diameter and 50 millimetre long or around 10,000 fuel particles are embedded in a graphite matrix to form a 60 millimetre TRISO fuel pebble. TRISO fuel is typically HALEU. BWX Technologies (United States) is licensed to produce HALEU and has the only current licence in the United States to produce Category 1 nuclear material including HEU. BWXT are restarting their TRISO fuel production line at Lynchburg, Virginia. It is possible that Australia could host a TRISO fuel facility, if prohibitions are lifted.

Australia (lead by ANSTO) is now a member of the GEN IV Forum (GIF) which is progressing the research and development for advanced reactor designs including the VHTR. Many of the six GEN IV designs have materials challenges and ANSTO's capability in this field is one of the reasons that Australia was invited to join the exclusive GIF club. Nuclear propulsion requires a compact design, typically a PWR. A gas-cooled reactor will always be physically larger than a water-cooled reactor so it is unlikely that a VHTR will be adopted for nuclear propulsion. Fuel fabrication is the last step of the 'front end of the nuclear fuel cycle'. The fuel is now ready to be loaded into a nuclear reactor.

The 'back end'

Generation of electricity from nuclear reactions produces waste. The three types of nuclear waste are classified according to their level of radioactivity: low, intermediate and high. Most of the waste—90 percent of the total volume—is composed of only lightly-contaminated items, such as tools and protective clothing, containing only one per cent of the total radioactivity. By contrast, high-level radioactive waste—mostly comprising used nuclear (sometimes referred to as 'spent') fuel accounts for just three per cent of the total volume but contains 95 per cent of the total radioactivity. The production of nuclear waste needs to be seen in context. The generation of electricity from a typical 1000-megawatt nuclear power station supplying the household needs of more than a million people produces only three cubic metres of vitrified high-level waste per year when the used fuel is recycled. In comparison, a 1,000-megawatt coal-fired power station produces approximately 300,000 tonnes of ash and more than six million tonnes of carbon dioxide, every year. Effective and efficient handling of waste is an integral part of the nuclear fuel cycle. When a reactor is refuelled, the used fuel is initially stored in a cooling pond close to the reactor. The water shields operators from radiation and also removes the decay heat from the fuel assembly.

New fuel in a power reactor typically contains four per cent U-235 and 96 per cent U-238. A typical PWR fuel assembly will stay in the reactor for 5–6 years, during which time the fissions in U-235 will have reduced its proportion to around one per cent (still higher than 0.7 per cent natural uranium). The fission products are the main source of heat and radioactivity initially. Most of the U-238 is unchanged, but neutrons captured in U-235 result in the production of actinides. Many of these have long half-lives and are the most significant contribution to decay heat and radioactivity in used fule after the first 100 years. A complete used fuel assembly will take more than 100,000 years to decay back to the level of the original uranium ore.

After storage in a cooling pond, there are four options for the management of used fuel: dry cask storage, reprocessing, final disposal and use as fuel in a fast reactor.

Dry cask storage

Particularly for countries that do not reprocess their used fuel, the complete fuel assemblies can be transferred to storage in a dry cask, usually on the reactor site, but can be at a central facility. The massive (typical 100Te) metal or concrete cask stores 24–90 fuel assemblies depending on type of fuel and burnup. Heat is passively removed by natural air circulation. The cask design has to take into account criticality, heat removal, radiation protection, containment against hazards and a 100-year lifetime. Major suppliers of casks are Transnuclear (France), NAC International (United States), Holtec (United States) and GNS (Germany). Dry cask storage is extensively used worldwide for interim storage of used fuel.

Reprocessing

Ninety six per cent of the nuclear material in a used fuel assembly can be recycled. Uranium and plutonium are recovered and can be used as reactor fuel. The waste volume is reduced, short and long term radiotoxicity is reduced by a factor of 10 and the waste form is suitable for long term storage.

Plutonium Uranium Extraction (PUREX) is a mature technology with high recovery yield—99.9 per cent of the uranium and 99.8 per cent of the plutonium. The disadvantages are the complex, expensive plant requiring careful control of contamination, radiation and criticality; the cost of reprocessing and proliferation concerns. Most used fuel can be reprocessed using PUREX technology. The fuel types that are difficult to reprocess are TRISO fuel and silicide fuel.

Around a third of fuel discharged has been reprocessed. There are PUREX plants in France, Japan, United Kingdom, Russia and India. The French COGEMA La Hague reprocessing facility is the largest in the world with a capacity of 1,700 tHM/year, and routinely reprocesses 1,150tHM/year from EdF power reactors in France.[4] Some countries, particularly the United States, Canada, Finland and Sweden, do not reprocess used fuel for policy reasons.

The PUREX process starts with the shearing of a fuel assembly into short sections which are then dissolved in hot nitric acid. The structural metal parts do not dissolve and are washed and collected and compacted into universal containers. Tri-butyl-phosphate (TBP) diluted with kerosene is added to the nitric acid aqueous solution, mixed and settled. The uranium and plutonium transfer to the organic solvent, the fission products and actinides stay in the aqueous solution. The uranium and plutonium are separated by adding a reducing agent, precipitating out and drying to produce the final products UO_2 and PuO_2. In the vitrification process, the fission products and actinides are melted with glass frit for several hours and then poured into a 180 litre universal container which is stored, before final disposal as High Level Waste (HLW).

Australian reprocessing policy

ANSTO's HIFAR research reactor operated from 1958–2007. There were 729 HIFAR fuel assemblies of US origin fuel that were returned to the United States under the Foreign Research Reactor Spent Nuclear Fuel (FRR SNF)

Acceptance Program, with no return of waste to Australia. There were a further 1288 HIFAR fuel assemblies with fuel not of American origin that were sent to France for reprocessing 1999–2004. This resulted in only five cubic metres of vitrified waste in 20 universal containers in one dry cask returning to Australia in 2015. This dry cask is currently stored in the Interim Waste Facility at Lucas Heights as Intermediate Level Waste (ILW). It is ILW because of the low heat load of the old research reactor fuel. The used fuel from ANSTO's OPAL research reactor is also being sent to France for reprocessing. The first shipment was despatched in 2018. It is unlikely that Australia would build sufficient power reactors to justify the cost of a reprocessing facility for its own use.

Final disposal of fuel assemblies

The IAEA standard for the final disposal of High Level Waste (HLW) is deep geological disposal.[5] Complete used fuel assemblies and the containers of vitrified waste from a reprocessing plant are HLW.

Three types of rocks are suitable as host sites for a geological disposal facility: high strength rocks like granite; low strength sedimentary rocks like clay; and, rock salt. The repository under construction in Finland at Onkalo is 455 metres deep in granite using the Swedish KBS-3 system. Twelve used fuel assemblies are loaded into a 4.8 metre long, copper canister, which is placed by a machine in a vertical hole from one of the underground galleries. The void between the bedrock and the canister is filled with bentonite clay which protects the can from corrosion and minor movement in the rock. Construction of the same type of repository at Forsmark in Sweden is about to commence. Cigéo (Centre Industrial de Stockage Géologique), the French deep geological repository, is 500 metres deep in clay. The site is near to Bure, 300 kilometres east of Paris and construction commenced in 2019. Switzerland is also examining sites in clay. The Waste Isolation Pilot Plant (WIPP) near Carlsbad in New Mexico has been in operation since 1999 taking transuranic waste from American military programs. The repository is 655 metres underground in rock salt. It does not accept civil HLW.

An alternative concept is deep borehole disposal, using vertical drilling to a depth of one kilometre or more followed by a gentle curve and one kilometre long horizontal drilling. The America company, Deep Isolation, is developing the safety case for this technology and, in January 2019, demonstrated placement and retrieval of a canister 600 metres underground.[6] A canister is typically 23–33 centimetres in diameter and 4.3 metres long and holds one fuel assembly. One 0.48 metre diameter borehole can store 11 years used fuel from a PWR. The cost is much less than a mined underground repository and there is no need for humans to go underground. Disposal sites could be near the nuclear power plant. Canada's Nuclear Waste Management Organisation (NWMO) has also drilled test boreholes. In Australia, the CSIRO, in partnership with ANSTO and Sandia National Laboratory (United States), is examining borehole technology.

The South Australia Nuclear Fuel Cycle Royal Commission (SANFCRC) considered the possibility of a deep geological repository in South Australia. The geology of the region is old and stable and has both hard crystalline rock and appropriate sedimentary formations including clay. The royal commission found that, globally, there are substantial quantities of used fuel in temporary storage awaiting final disposal. The concept was to take used fuel from overseas, interim store in dry casks with final disposal in a deep geological repository. Financial analysis showed a very large economic benefit to South Australia.

Final disposal option—use as fuel for a fast reactor

Used fuel from a thermal power reactor contains uranium, plutonium and other actinides.[7] Actinides that have a negligible probability of fission in a thermal reactor will fission in a fast reactor, that is, a reactor without a moderator to slow the neutrons. Used fuel from thermal reactors can, therefore, be used as fuel for a fast reactor, leaving less long-lived waste to be managed. This process was demonstrated in the EBR II sodium-cooled *Integral Fast Reactor* project in the United States (1963–94) along with associated pyroprocessing on site. The six reactor types being developed by the Gen IV Forum (GIF) include three types of fast reactor (sodium cooled, lead cooled and gas cooled) and also the molten salt cooled reactor that can be fast or epithermal. One of the benefits of these fast reactors is the consumption of actinides from particularly the current light water reactors, so reducing the amount of used fuel for final disposal.

The management of other radioactive waste

In addition to used fuel, there are other types of radioactive waste produced from the nuclear fuel cycle, including waste from uranium mining. The IAEA Classification of Radioactive Waste lists the different categories of waste and the best practice final disposal technology.[8] The operation of nuclear reactors, including reactors for nuclear propulsion, produces Low Level Waste (LLW) and Intermediate Level Waste (ILW). The only High Level Waste (HLW) is the used fuel as managed in the previous section. In addition to the radiological characteristics of radioactive waste, the physical and chemical properties must also be considered.

Low level waste (LLW)

Gaseous radioactive waste is filtered and allowed to decay before discharge within limits set by the nuclear regulator. Liquid radioactive waste containing low radioactivity solids is filtered, treated by typically ion exchange and discharged within limits. If high radioactivity it is solidified for disposal. Low level solid waste is typically cleaning materials, clothing that cannot be laundered and filter cartridges. The waste is compacted into steel drums. A typical large reactor will produce around 120 cubic metres annually of packaged LLW which would fill two shipping containers. International best practice for the final disposal of LLW is a near surface repository which is an engineered facility with multiple barriers. A good example is the El Cabril Disposal Facility in Spain.[9]

The Australian Commonwealth has accumulated around 5000 cubic metres (including packaging) of LLW, comprising around 2000 cubic metres from over 60 years of ANSTO's operations at Lucas Heights and 2100 cubic metres from research into processing of radioactive ores held by CSIRO at Woomera, in addition to small quantities held by other organisations. Current production of LLW is less than 40 cubic metres annually. In January 2020 following a voluntary process, the Department of Industry, Science, Energy and Resources announced that Napandee in South Australia is the preferred site for the Australian National Radioactive Waste Management Facility.[10] The facility will be for the permanent disposal of LLW and the temporary storage of ILW.

Intermediate level waste (ILW)

Operation of a nuclear power plant produces small amounts of ILW, typically 1.5 cubic metres annually, mainly metallic waste from maintenance and refuelling operations. International best practice for final disposal of ILW is underground at an intermediate depth. An example is the SFR facility at Forsmark in Sweden, which is 60 metres underground.[11] The Australian Commonwealth inventory has a total of 656 cubic metres of ILW, mainly from research reactor operations and radiopharmaceutical production. Current production is less than five cubic metres annually. Plans for the final disposal facility for Australian ILW are still being developed.

Conclusion

Australia has the world's largest reserves of uranium and is the world's third largest producer but all the uranium is exported for processing overseas. At present the conversion, enrichment and fuel fabrication processes are dominated by a small number of big players and there is over-capacity in the market. There may be opportunities for Australia in the future, particularly in fuel and materials for advanced reactors, but the prohibitions on nuclear activities in Australia have to be removed first. Australia has the potential to be a major player in the back end of the nuclear fuel cycle (management of used fuel). As identified by the SANFCRC, there is a global need and there would be very substantial economic benefits to Australia by providing a used fuel management service. This would require a Government with vision, prepared to look at innovative ideas, with long-term policy objectives.

Endnotes

1 Final report of the South Australia Nuclear Fuel Cycle Royal Commission https://yoursay.sa.gov.au/pages/nuclear-fuel-cycle-royal-commission-report-release/.

2 SWU calculations: https://www.urenco.com/swu-calculator.

3 URENCO sites and technology https://www.urenco.com/.

4 tHM/year = tonnes of heavy metal/year—the nuclear material part of a fuel assembly.

5 IAEA Safety Standard Geological Disposal Facilities for Radioactive Waste SSG-14 https://www.iaea.org/publications/8535/geological-disposal-facilities-for-radioactive-waste.

6 Deep Isolation (USA) borehole technology https://www.deepisolation.com/technology/.

7 A *thermal* power reactor has a moderator to slow the neutrons down to thermal energies to improve the probability of fission in U-235, Pu-239 and Pu-241.

8 IAEA Classification of Radioactive Waste GSG-1 https://www.iaea.org/publications/8154/classification-of-radioactive-waste.

9 El Cabril Disposal Facility http://www.enresa.es/eng/index/activities-and-projects/el-cabril.

10 Australian National radioactive waste Management facility https://www.industry.gov.au/strategies-for-the-future/managing-radioactive-waste.

11 SFR final repository for low and intermediate level radioactive waste in Sweden https://www.skb.com/our-operations/sfr/.

6 Nuclear power for electricity

Ben Heard

Energy underpins human civilisation. It is 'the only universal currency'[1] and 'the master enabler'.[2] Civilisation has been a transitional process of energy exploitation: from using sticks and stones as tools and weapons to the point where 'today, there are none living without the products of agriculture and large-scale industry'.[3]

This transitional process of increasing energy exploitation has been good for humans. Since around 1800, average life expectancy has risen from 30–40 years to 71.5 years in 2015.[4] The world is now less violent than at any other point in our history[5] and global vaccination coverage of newborn children is now 86 per cent.[6] The global rate of population growth is now 1.05 per cent , continuing its decline since peaks of approximately 2–2.3 per cent around 1963–1970.[7] Energy underpins global food security, sanitation, modern medicine, and education, and these historical trends point to energy security as an enabler of peace, technological advancement and human longevity.

Then, there is electricity. By learning to create and control a flow of electrical charge, we have provided ourselves with a versatile energy form which, in little more than a century, transformed the face of the modern world. Application of electricity has been diverse and transformative. The most emblematic invention, the humble incandescent bulb, used simple electrical resistance to convert current to visible light, spelling the end of candles and kerosene lamps. Less well-remembered, but no less transformative, electric motors brought a new scale of locomotion to society, powering early public transport in the form of trolley cars, and driving the elevators of the skyscrapers that changed the face of our cities.

Yet, just under a billion people world-wide still lack access to electricity—defined as an intermittent supply of up to 1,250 kWh per year for an urban household.[8] That is an order of magnitude less than consumption in Australian households.[9] Reliable, stable power supply also underpins the economic growth and industrialisation that enables transition from subsistence agriculture towards the modern economies that can deliver the great advances in human well-being. For health, security and safety, virtually everyone in the world wants electricity. So, in seeking to create a stable, sustainable, equitable world, we must think carefully about the relative merits of different ways of creating electrical power. Currently, around 75 per cent of global electricity is provided by thermal generation—using heat to make steam for spinning turbines.

Among so many primary energy sources, nuclear fission brings a range of important advantages. Nuclear power:

- is low-carbon, boasting low life-cycle emissions
- has among the smallest land and resource footprints of any energy source
- helps to avoid pollution such as NOx, SOx, heavy metals and particulate matter
- is flexible in terms of siting and not limited by fuel resource availability and transport infrastructure (railways, pipelines, etc)
- is capable of operating reliably at capacity factors in excess of 90 per cent, but can also load follow if desired[10]
- is long-lived infrastructure. Plants commissioned today have design lives of 60 years and might still be providing the world with power at the turn of the 21st century
- provides rotational inertia that helps to stabilise the grid and regulate frequency
- allows stockpiling of fuel, which boosts security of energy supplies
- Is among the most cost-competitive energy options over many decades of operation[11]

- is a major employer in rural areas, supporting skilled hi-tech jobs and local economic activity
- offers proven technology that is available today and can be scaled comparatively quickly[12]
- can provide isotopes and support for research, medicine, industrial and agricultural purposes
- is improving, with new technologies offering greater efficiencies and opening up new applications to enable decarbonisation of heat, industry and transport sectors

Perhaps the most salient of advantages in the current discussions on energy relates to high reliability delivered with such a low environmental footprint. In nuclear fission, humanity has fuels that *do not burn*, meaning no emissions of carbon dioxide—or anything else. So, while providing a decisive technological response to mitigating climate change, nuclear power stations also avoid and displace the dangerous air-pollutants that, under fossil and biomass driven systems, are externalised to the environment. These and other environmental advantages are discussed in detail in a subsequent chapter.

Power production from nuclear fission also brings challenges. To date, nuclear power plants tend to be very large projects that test manufacturing and project delivery capabilities. Building and operating a nuclear power sector demands a nation attains, and maintains, institutional capacity and capability in science, regulation, and engineering. Nuclear plants also produce highly radioactive used nuclear fuel. This is small in volume, but presents specific waste management and disposal problems.

With these advantages, fission has achieved and maintained a substantial share of global power production. However, for an accurate understanding of its place and prospects, the data needs to be considered from several perspectives. In 2019 nuclear technologies provided 2796 TWh of electricity world-wide, accounting for 10.4 per cent of global supply.[13] This is close to the maximum historic supply (2807 TWh in 2006), having recovered nearly all the supply that suddenly exited in Japan following the nuclear accident of 2011. Total production has grown every year from 2012. Share of global supply has, however, fallen from a peak 17.5 per cent in 1995.[14] That falling share can be traced to several trends.

First, global electricity consumption doubled between 1996 and 2018.[15] Most of this growth was delivered by increased consumption of coal and gas, with large increases in consumption in many developing nations where nuclear sectors have not yet been established.

Second, markets in which nuclear power was already established, particularly Western Europe and the United States, had a sustained hiatus of around 20 years in new construction.[16] Such construction hiatus has been identified as a strong driver of cost escalation of nuclear power development in Western nations, where skills, experience, and regulatory capacity have declined and must be re-established. Where the United States and Western Europe have built new power stations, the hiatus of recent decades has proven costly and difficult to overcome. There have been price and construction time overruns in the United States,[17] France, Finland and the United Kingdom.[18]

Third, many nuclear assets have closed during this time, due to inextricably linked political and economic pressures. New build is substantially offset by loss of supply elsewhere. The clearest examples of premature closure is that of Germany, where 12.3 GWe of capacity was retired from 2011–18,[19] and Japan where 18 operable plants await sufficient support for restarting after achieving regulatory approval following the Fukushima accident of March 2011 (discussed below).[20] Several examples of premature closure can also be found in the United States, including Vermont Yankee[21] and Indian Point,[22] as well as South Korea and Taiwan. Nuclear plants, when well-managed and maintained, have been shown to deliver substantially greater output in the course of their operating lives as experience develops and refurbishment is undertaken. For example, the United States has seen increased total output from nuclear sources despite an overall net loss of installed capacity, driven by operational improvements delivering substantially greater output, or 'uprating' from the same assets.[23] With no fixed technical limit on the lifespan of a nuclear reactor, and some reactors now achieving 50 years of operation, it can be argued that all nuclear closures to date have been premature losses of zero-carbon generating capacity,[24] at a time when the world can ill-afford such losses.

Fourth, growth in nuclear power production since 2012 has been substantially concentrated in China—responsible for 86 per cent of the growth in annual supply from 2012–2018. With this sustained program of construction, building new nuclear plants in China has become cost-effective and completed on schedule, with recently announced plans to continue building 6–8 reactors per year, and raise total capacity to around 200 GW by 2035.[25]

To summarise, nuclear power remains a cornerstone of clean energy supply, both globally and in several major nations. With many crucial technical and environmental advantages, its percentage share of global power has recently stabilised and is increasing slowly. With fast growth in total electricity consumption, and so much of that growth occurring in non-nuclear, developing nations, the percentage share from nuclear technologies is, however, under pressure. Interrelated political and financial challenges affect efforts to build new assets, and also retain existing assets in an ageing fleet. For nuclear technologies to avoid becoming a niche contributor to power, the 10 years to 2030 will need to look very different, and that particularly will require accelerated efforts to broaden the use of nuclear power around the world, with swifter introduction and growth in emerging economies.

There are reasons to believe that will be the case. The number of reactors under construction, proposed and planned suggests the recent growth in total generation is likely to be maintained with a strong pipeline of projects. While new build is dominated by China (12.2 GWe under construction), a further 50 GWe of new building is underway in 17 nations. The new entrants include Bangladesh, Belarus, Turkey, and the United Arab Emirates (UAE). Building in Finland, the United Kingdom, France and the United States, is plagued with delay but nevertheless progressing with some nearing completion.[26] This will bring approximately 10 GWe online. The Barakah nuclear power plant in the UAE, has commenced commissioning in 2020,[27] bringing 4.6 GWe of capacity online swiftly and at an attractive price. Several potential newcomer nations in Asia and Africa are working with the International Atomic Energy Agency and other national partners to develop institutional capacity and readiness for adopting nuclear power.[28]

This progress remains hard fought. Institutional prejudices and barriers exist in both national policies and multinational institutions. Australia is a prime example, where environmental legislation renders approval of a nuclear power project impossible, despite its environmental merits.[29] Nuclear technologies were excluded from consideration under the Clean Development Mechanism of the Intergovernmental Panel on Climate Change,[30] meaning developed nations would receive no recognition for delivering a nuclear power development in place of more polluting power stations. The World Bank 'doesn't do' nuclear energy,[31] and similarly the Asian Infrastructure Investment Bank will not finance nuclear projects, but will support fossil fuel based generation facilities based on 'commercially available least carbon technology'.[32]

The publications of the IPCC and the IEA[33] have, however, both become more forthright that nuclear technologies will be an essential part of transitions to lower carbon economies. The most aggressive mitigation scenarios published by the IPCC indicate a greater than five-fold increase in generation from nuclear power in the 30 years to 2050.[34] This adjustment in posture comes after thirty years during which the overall global reliance on fossil fuels remained unchanged, despite impressive progress in the electricity sectors of some jurisdicitons.[35]

The world has reached an important juncture for nuclear technologies. There is an increasing public and institutional consensus that there will need to be a greater role for nuclear technologies in achieving a successful low carbon energy transition. Yet a plethora of multinational, national, and sub-national institutional barriers remain. New nuclear nations are poised to come online, and new nuclear technologies have the potential to break out in global markets. But there remains attrition in nuclear fleets that is presently only just outmatched by new building. For nuclear to play a greater role, the premature retirement of nuclear plants needs to stop, and the barriers to new building in emerging economies and newcomer countries need to be removed.

Like perhaps no other technology, the question of safety remain a millstone around the neck of nuclear power. Consideration of nuclear safety might be broadly split into two categories: statistics and stories. For policy-makers it is well-worth being acquainted with the important lessons of both. Beginning with the statistics, nuclear power has, during its 60 year history, been an exceptionally safe way of generating electricity. Studies of mortality and morbidity make a convincing case that the popular reputation of nuclear is back to front. Instead of its being an exceptional hazard, we have actually caused, and continue to cause, great harm failing to use enough of it.[36] Nuclear fission requires less from the most dangerous parts of energy—mining vast quantities of coal, transporting huge quantities of flammable gas or oil, or emitting hazardous pollution. Nuclear gives rise to a small mining footprint, increasingly using the minimal-disturbance technique of in-situ recovery.[37] Its fuel is a heavy, stable solid, neither flammable nor explosive.

Even before taking into account regulation and an advanced safety culture, nuclear power presents a comparatively safe and healthy option for electricity supply. Between 1978 and 2008, nuclear power in the all EU-27 and OECD nations recorded no accidents in excess of five fatalities, where oil recorded 250 such accidents, causing 4500 deaths in the same period.[38] We appear sadly inured to the banal death and illness from familiar, everyday sources. It is reminiscent of the old joke where, having announced the sudden death of a loved one, a friend asked what they died from. The answer? 'Nothing serious.'

Serious accidents can, do, and have occurred in the use of nuclear technologies—as is true of all technologies. But our response has constrained the growth of nuclear technologies in ways that have simply ushered greater dependence on vastly more dangerous energy systems. For policy-makers, both the accidents and the responses are worth knowing and understanding in order to put coming decisions into an adequate context. The explosion and subsequent fires on 26 April 1986 at the Chernobyl nuclear power station in Ukraine (then part of the Soviet Union) dominates public perceptions of the inherent hazard of nuclear. I was aged seven at the time of the accident, and images of this incident from *National Geographic* were formative in my development as a hard anti-nuclear environmentalist. This accident is an appalling outlier. A fulsome coverage is beyond the scope of this chapter although some aspects are so consequential that Australian policy-makers ought to appreciate the reasons why an accident of this type could not be repeated.[39]

The reactor in question was a *reaktor bolshoy moshchnosty kanalny*, or RBMK, a design from the mid-1960s which was dual-purpose, suitable for making weapons grade plutonium at the same time as producing electricity. This design requirement drove several design features, including the combination of a graphite moderator with normal water as the coolant. No other reactor features that combination, and no such design has been licensed by Western regulators. If operated well-outside normal conditions, the reactor had the potential to experience strong, interrelated feedbacks that could rapidly increase the power in the reactor core beyond safe conditions, overwhelming safety systems and creating a steam explosion. At Chernobyl nuclear power station, the reactor was placed in absolutely *abnormal,* low-power operations as part of an engineering experiment, to determine whether water pumps to the reactor might be operated by turbine inertia in the event of power failure.

Under these abnormal conditions, including manipulation of the water circulation systems, the feedbacks described above came into play as the operators ignored established practice. In the end, the power levels of the reactor spiked 100 times beyond its design in a matter of seconds. The reactor core exploded, and the reactor caught fire. The graphite core (being pure carbon) burned with damaged fuel rods inside it. The RMBK design lacked *robust containment,* as is standard in other commercial reactor designs. When the reactor exploded, there was nothing between it and the environment other than a small amount of concrete shielding. Vapourised radioactive material was distributed in a plume—the most potent pathway for distributing radioactive material.

Comprehensive investigation by the United Nations Scientific Committee on the Effects of Atomic Radiation (UNSCEAR) describe that as well as two fatalities relating to the initial explosion, a further 134 workers experienced radiation sickness, and 28 received such large doses of radiation they died from acute radiation syndrome.[40] The release of radioactive material, including the common fission product iodine-131, resulted in an excess in thyroid cancer in members of the general public. A substantial fraction of approximately 6000 cases was considered attributable to radiation exposure up to 2005. Fortunately, there were only 15 fatalities from this group.[41] Beyond this, the UNSCEAR reported there has been 'no persuasive evidence of any other health effect in the general population that can be attributed to radiation exposure', that 'the vast majority of the population need not live in fear of serious health consequences from the Chernobyl accident', and that 'from the radiological point of view, generally positive prospects for the future health of most individuals involved should prevail.'[42] Updated estimates affirm these general results.[43]

Yet other fatality data for this accident include estimates in the tens of thousands[44] to up to a million people.[45] Generating these findings usually relies on measuring the collective dose of a population, and applying the linear no-threshold (LNT) model that assumes any exposure to radiation entails a degree of harm. Both UNSCEAR and the International Council for Radiological Protection expressly recommend against using that approach to infer harm after the fact.[46] The ICRP states:

> the computation of cancer deaths based on collective effective doses involving trivial exposures to large populations is not reasonable and should be avoided. Such computations based on collective effective dose were never intended, are biologically and statistically very uncertain, presuppose a number of caveats that tend not to be repeated when estimates are quoted out of context, and are an incorrect use of this protection quantity.

There seems a persistent, ghoulish need to attribute more death to nuclear technologies, instead of welcoming the good news that this energy source with such a low environmental impact *also* improves human health and safety. Given the World Health Organisation estimates an *annual* toll of three million premature deaths from ambient air pollution,[47] we appear to be poor at measuring our monsters and deciding which ones to fight. A better understanding of the actual, as opposed to popular, impacts from low level radiation exposure, lets us use knowledge to prevent the widespread harm that continues from our energy choices today.

The accident at Fukushima Daiichi nuclear power station in March 2011 occurred in a reactor design that *was* licensed in the West—Mark 1 Boiling Water Reactors were designed in the 1960s and commissioned from 1971. In this reactor design, decay heat is managed through pumped circulation of water. With the loss of offsite power due to the massive quake, the pumping was dependent on diesel generators to manage the decay heat in the core, after the reactors had safely shut down. In a now-obvious error, the generators for several of the reactors were located in basement facilities. These facilities were inundated when the massive tsunami overwhelmed the sea-wall protection. Power to the plants was lost. As a result, so was the capacity to cool the core.

The fuel cores of three of the reactors melted. As water in the reactor core turned to steam, this steam interacted with the steel interior and created a build-up of hydrogen gas. This gas vented from the core and the containment into the space between the containment and the outer building. Here, venting failed, and hydrogen explosions occurred, causing significant damage to the structures. Radioactive material was released and drifted predominantly in a north-west direction.

The more recent reactor units, with elevated generators for back-up power, withstood the event without damage to the core.

Just days before this accident occurred and before an audience of 100 people, I declared that I changed my mind and had become a strong supporter of nuclear power. That weekend I watched in awe and horror as that black wave engulfed Japan, and news of a stricken nuclear plant began to hit the headlines. I have since visited the site of Fukushima Daiichi three times. Most recently I toured the region to record the residual radiation (finding it to be predominantly normal) and entered the damaged reactor unit 2 with an Australian television crew. We discussed the event and the consequences. There have been no direct fatalities as a result of the radiation release from this accident and there is never expected to be any detectable health impact as a result of radiation exposure. My conviction is that this accident needs to lead us towards better safety. It must not lead us away from essential decarbonising technology.

The global nuclear power industry enacted a powerful response. In the United States, where Mark 1 BWR reactors are in use, extensive additional mitigations and protections were installed at nuclear power plants throughout the country, including permanently installed back-up power, portable on-site back-up power, and a national rapid response unit of emergency equipment.[48] Additional hardened venting was installed in the Mark 1 BWR plants, along with new sensor equipment in the used fuel pools. The European Union enacted a massive multilateral peer review process[49] to revise the adequacy of safety systems in nuclear power stations in all member countries. Like all industries, nuclear power is imperfect but the safety culture is decidedly strong. A failure anywhere tends to be addressed as a responsibility everywhere.

The traditional narrative is that nuclear accidents are the quintessential 'low-likelihood, high consequence' industrial accident. An ironic result of the accident at Fukushima Daiichi has been to demonstrate that, in contrast with other power sources, this is actually incorrect. Nuclear accidents are low-probability, medium-consequence, provided we keep our responses in proportion to the underlying hazard. The renowned environmentalist, George Monbiot, probably said it best in the weeks following the accident:

> A crappy old plant with inadequate safety features was hit by a monster earthquake and a vast tsunami. The electricity supply failed, knocking out the cooling system. The reactors began to explode and melt down. The disaster exposed a familiar legacy of poor design and corner-cutting. Yet, as far as we know, no one has yet received a lethal dose of radiation ... Atomic energy has just been subjected to one of the harshest of possible tests, and the impact on people and the planet has been small. The crisis at Fukushima has converted me to the cause of nuclear power.[50]

Taking Monbiot's perspective to the present day, Australia cannot acquire any design from 40–60 years ago. We can only acquire designs that have addressed these problems and defects in the most basic aspects of design and planning, to be operated by a global safety culture that operates under expectations of continuous learning and improvement.

As much as one might wish to be steered by statistics, stories are powerful. It can be difficult to shake the sense that nuclear brings *exceptional* danger. In truth, stories of harm from other energy sources go largely untold. It is a sad experience to seek comparative accidents in energy that no one has heard of; accidents that did not prompt any global outcry or action. It is sad because there are so many to choose from, and they are so much worse.

- Two years before Chernobyl, the LPG gas explosions of San Juanico, Mexico, killed 500–600 people and injured more than 7,000.
- Three years after Fukushima, a fire in a coal mine in Soma, Turkey, killed 301 miners of burns and suffocation.
- In 2015, an explosion at a gas station in Accra, Ghana, killed over 250 people.
- Four months before Fukushima, an explosion at the Pike River Coal Mine in New Zealand killed 29 people.

At the time of writing (2020), an explosion at a coal mine in central Queensland has left five miners seriously injured. Over coming decades,

Australia faces a transition of its energy system. Thanks to advances in technology, and Australia's existing blank slate in nuclear power, as a new entrant Australia has the opportunity to make its energy system profoundly safer using the newest and best of 21st century nuclear power.

Australia can position itself to be a fast-following adopter of small modular reactors (SMRs). Small modular reactors are advanced reactors that produce up to 300 MWe per module, with many expected to be made in factories, and shipped to the destination power plant by road or rail.[51] Most of the SMRs under development have incorporated advanced passive or even inherent safety features. Some SMRs are targeted at single module developments, with others intended to expand in line with demand to make up multi-module power plants.[52] By moving from a construction to a manufacturing paradigm, and with designs that have been conceived at the outset with lower costs in mind, SMR power plants are intended to be easier to finance and build, and more reliably lower-cost than very large plants. With smaller generating units, they are also better suited to jurisdictions where the electricity grid itself remains an underlying constraint that makes connections of single units of 1000 MWe or more challenging without substantially enhanced infrastructure.

Advanced nuclear developments have received sustained bi-partisan policy and licensing support in the United States. The world's largest user of nuclear power might soon recover advantage it has ceded in the large reactor space. The 60 MWe unit from NuScale Power can combine 12 modules for a 720 MWe power plant.[53] Each unit is a self-contained reactor core and steam generator, to be connected to a dedicated 60 MWe turbine. This reactor recently became the first SMR to be licensed with the Nuclear Regulatory Commission of the United States,[54] with a first project slated for power production in 2029. The NuScale Power Module has an innovative core cooling function. The modules operate in a massive heat sink—a reservoir of water. In the event of a total loss of offsite power, as occurred in Fukushima, safety valves open to circulate heat to this heat sink. The reactors remain in a safe condition indefinitely—no external power is required to prevent core damage. The safety case is sufficiently compelling that the United States Nuclear Regulatory Commission has issued a proposed rule change, open to comment at the time of writing, to reduce the emergency preparedness zone to the site boundary.

Another important example in advanced nuclear is the 195 MWe Integral Molten Salt Reactor from Canadian developer, Terrestrial Energy. This reactor, which is in advanced stages of pre-licensing vendor design review in Canada, offers a distinct advantage of higher operating temperatures of 600 degrees celsius. This provides a compelling option for the decarbonisation of industrial heat and the production of synthetic fuels and hydrogen, as well as the production of affordable and reliable electricity.

The use of a liquid fuel brings a string of interrelated cost and safety advantages. 'Meltdown' as a concept is redundant—the fuel is molten in normal operations. If the reactor loses criticality and loses power, the fuel would gradually cool and solidify. The reactor has no water in the core, so operates at atmospheric pressure, and there can be no production of hydrogen gas. Finally, in the event of any inadvertent operations at higher temperature, the physics of the liquid fuel ensures it reduces in criticality and slows the chain reaction. It simply cannot be pushed to operate above its design, rendering it 'walk-away safe'.

The future of nuclear technology remains open. Between the pipeline of 55 reactors under construction with more than 100 planned, and the tangible progress in commercialisation of advanced reactors, there are many reasons to believe the ten years to 2030 will see sustained growth in nuclear power production world-wide. Nuclear technologies have made a major contribution to reliable, affordable low carbon electricity supply through the 20th century and into the 21st century. As the imperative to both decarbonise energy supplies and replace ageing infrastructure becomes ever-greater, a sustained and growing role for nuclear technologies might appear logical and assured.

The experience of the previous three decades proves otherwise, however—the development of clean energy economies is not assured. It is a choice nations must make, and a journey of decades, to reach a new operating state. Full-decarbonisation is a condition that must be sustained in-perpetuity in order to retain a stable climate.

Given Australia's comparatively high dependence on fossil fuels, our journey will be more fraught than for most nations. To ignore an entire family of assistive technologies runs counter to common sense. Australia could now choose to position itself to take full advantages of these technologies, in support of urgent re-invention of our power system, building on our existing foundations in science, technology and engineering, and deliver a healthier, cleaner, and safer energy future for generations to come.

Endnotes

1   V Smil, *Energy and Society: A History*, 2017, Massachusetts, MIT Press.

2   CAS Hall, *Energy Return on Investment A Unifying Principle for Biology, Economics, and Sustainability*, Lecture Notes in Energy, vol. 36, 2017, Cham, Springer.

3   J Goudsblom, *Energy and Civilization.* International Review of Sociology, 2012, vol. 22, no. 3, pp. 405-411.

4   World Health Organisation. *Global Health Observatory (GHO) Data - Life expectancy.* 2017: http://www.who.int/gho/mortality_burden_disease/life_tables/en/. Gapminder. *Gapminder data - life expectancy at birth, with projections*, 2017: http://www.gapminder.org/data/.

5   S Pinker, *The Better Angels of Our Nature.* 2011, New York, Viking.

6   World Health Organisation. *Immunization coverage.* 2017 July.

7   Worldometers. *World Population.* 2017: http://www.worldometers.info/world-population/.

8   International Energy Agency, *World Energy Model Documentation.* 2019: Paris.

9   https://ahd.csiro.au/other-data/typical-house-energy-use/.

10   Nuclear Energy Institute, *Nuclear By The Numbers.* 2020: Washington DC.

11   OECD/IEA/NEA, *Projected Costs of Generating Electricity.* 2015, Organisation for Economic Co-operation and Development/International Energy Agency: Paris.

12   SA Qvist and BW Brook, *Potential for worldwide displacement of fossil-fuel electricity by nuclear energy in three decades based on extrapolation of regional deployment data.* PLoS One, 2015, vol. 10, no. 5, p. e0124074.

13   BP, *BP Statistical Review of World Energy.* 2019, London.

14   BP, *BP Statistical Review of World Energy.* 2019, London.

15   BP, *BP Statistical Review of World Energy.* 2019, London.

16   JR Lovering, T Nordhaus and A Yip, *Apples and oranges: Comparing nuclear construction costs across nations, time periods, and technologies.* Energy Policy, 2017, no. 102, pp. 650-654.

17   Josh Freed, et al. *Is nuclear too innovative?*, 23 February 2017: https://medium.com/third-way/is-nuclear-too-innovative-a14fb4fef41a.

18   H Vella, *Does the European Pressurised Reactor have a future?* . 5 December 2019: https://www.power-technology.com/features/does-the-european-pressurised-reactor-have-a-future/.

19 World Nuclear Association. *Nuclear Power in Germany*. December 2019: https://www.world-nuclear.org/information-library/country-profiles/countries-g-n/germany.aspx.

20 World Nuclear Association. *Nuclear Power in Japan*. 2020 March 2020: http://www.world-nuclear.org/info/Country-Profiles/Countries-G-N/Japan/.

21 Nuclear Energy Institute. *NEI: Vermont Yankee Closing a Great Loss*. 29 December 2014: https://www.globenewswire.com/news-release/2014/12/29/694177/10113619/en/NEI-Vermont-Yankee-Closing-a-Great-Loss.html.

22 R Bryce, *Indian Point nuclear-reactor shutdown a huge blow to New York's environment*. 2020 April 29: https://nypost.com/2020/04/29/indian-point-nuclear-reactor-shutdown-a-huge-blow-to-nys-environment/.

23 M Scott and O Comstock. *Despite closures, U.S. nuclear electrcity generation in 2018 surpassed its previous peak*: https://www.eia.gov/todayinenergy/detail.php?id=38792.

24 World Nuclear Association, *The Enduring Value of Nuclear Energy Assets*. 2020: London.

25 D Stanway, *China to build 6-8 nuclear reactors a year from 2020-2025*. 2020 July 9 [cited 2020 22 July]; Available from: https://in.reuters.com/article/china-nuclearpower/china-to-build-6-8-nuclear-reactors-a-year-from-2020-2025-report-idINKBN24A0DL.

26 World Nuclear News. *TVO seeks go-ahead for Olkiluoto EPR fuel loading*. 2020 9 April [cited 2020 12 May]; Available from: https://www.world-nuclear-news.org/Articles/TVO-seeks-permission-for-Olkiluoto-EPR-fuel-loadin.

27 Nuclear Engineering International. *Fuel loading completed at unit 1 of Barakah nuclear power plant in the UAE*. 2020 3 March [cited 2020 12 May]; Available from: https://www.neimagazine.com/news/newsfuel-loading-completed-at-unit-1-of-barakah-nuclear-power-plant-in-the-uae-7805266.

28 Thomson Reuters Foundation. *As Africa looks for clean power, interest in nuclear grows*. 2020 April 10 [cited 2020 12 May]; Available from: https://www.japantimes.co.jp/news/2020/04/10/business/africa-clean-power-nuclear/#.Xro3QRMzZhE.

29 BP Heard, *That day in December: the story of nuclear prohibition in Australia*, in *Decarbonise SA*. 2012, Adelaide.

30 International Atomic Energy Agency, *Nuclear Power and Market Mechanisms under the Paris Agreement*. 2017, Vienna.

31 World Nuclear Association. *World Bank should support all clean energy technologies*. 2017 [cited 2020 12 May]; Available from: https://www.world-nuclear.org/press/press-statements/world-bank-should-support-all-clean-energy-technol.aspx.

32 Asian Infrastructure Investment Bank, *Energy Sector Strategy: Sustainable Energy for Asia*. 2018.

33 International Energy Agency, *Nuclear Power in a Clean Energy System*. 2019: Paris.

34 Intergovernmental Panel on Climate Change, *Global Warming of 1.5 °C*. 2018, IPCC: Incheon.

35 BP Heard, *Clean,. Reliable. Affordable. The Role of Nuclear Technology in Meeting the Challenge of Low Greenhouse Gas Electricity Supply in the 21st Century*, in *School of Biological Sciences*. 2017, University of Adelaide: Adelaide.

36 A Markandya and P Wilkinson, *Electricity generation and health*. The Lancet, 2007. 370(9591): p. 979-990.

37 BP Heard, *Environmental impacts of uranium mining in Australia: History, progress and current practice*. 2017, Minerals Council of Australia: Canberra.

38 P Burgherr and S Hirschberg, *Comparitive risk assessment of severe accidents in the energy sector*, in *International Disaster and Risk Conference*. 2008, Paul Scherrer Institute: Davos, Switzerland; P Burgherr and S. Hirschberg, *Severe accident risks in fossil energy chains: A comparative analysis*. Energy, 2008, vol. 33, no. 4, pp. 538-553.

39 For more detailed discussion see Bernard L Cohen, *The Nuclear Energy Option*.

40 United Nations Scientific Committee on the Effects of Atomic Radiation, *ANNEX D HEALTH EFFECTS DUE TO RADIATION FROM THE CHERNOBYL ACCIDENT*. 2008: New York, NY.

41 United Nations Scientific Committee on the Effects of Atomic Radiation, *ANNEX D HEALTH EFFECTS DUE TO RADIATION FROM THE CHERNOBYL ACCIDENT*. 2008: New York, NY.

42 United Nations Scientific Committee on the Effects of Atomic Radiation, *ANNEX D HEALTH EFFECTS DUE TO RADIATION FROM THE CHERNOBYL ACCIDENT*. 2008: New York, NY.

43 United Nations Scientific Committee on the Effects of Atomic Radiation, *Evaluation of Data on Thyroid Cancer in Regions Affected by the Chernobyl Accident*. 2017: New York NY.

44 Greenpeace. *Chernobyl death toll grossly underestimated*, 2006: https://www.greenpeace.org/archive-international/en/news/features/chernobyl-deaths-180406/.

45 B Sweet, *One Million Chernobyl Fatalities?* 18 May 2010]: https://spectrum.ieee.org/energywise/energy/environment/one-million-chernobyl-fatalities.

46 JJ Cardarelli and BA Ulsh, *It Is Time to Move Beyond the Linear No-Threshold Theory for Low-Dose Radiation Protection*. Dose Response, 2018, vol. 16, no. 4, p. 1559325818779651.

47 World Health Organization, *Ambient air pollution: a global assessment of exposure and burden of disease*. 2016: Geneva.

48 United States Nuclear Regulatory Commission, *Safety Enhancements After Fukushima*. 2019: Washington DC.

49 ENSR Group, *Peer review report - Stress tests performed on European nuclear power plants*. 2012: Brussels.

50 G Monbiot, *Why Fukushima made me stop worrying and love nuclear power*, in *The Guardian*. 2011: London.

51 International Atomic Energy Agency. *Small Modular Reactor (SMR) Regulator's Forum*, 2020: https://www.iaea.org/topics/small-modular-reactors/smr-regulators-forum.

52 International Atomic Energy Agency, *Advances in Small Modular Reactor Technology Developments. A Supplement to: IAEA Advanced Reactors Information System.* 2018: Vienna, Austria.

53 NuScale Power. *Technology Overview.* 2020 [cited 2020 20 February]; Available from: https://www.nuscalepower.com/technology/technology-overview.

54 United States Nuclear Regulatory Commission. *Design Certification Application - NuScale.* 2017 October 24 [cited 2018 31 January]; Available from: https://www.nrc.gov/reactors/new-reactors/design-cert/nuscale.html.

7   Small modular reactors

Tony Irwin

*T*here is rapidly growing interest in Small Modular Reactors (SMRs) to provide electricity generation and other services to small grid systems, remote areas and industry. An SMR is defined by the International Atomic Energy Agency (IAEA) as a reactor with a power output of less than 300 MWe (1 MWe is one million watts of electric capacity), but typically they are smaller. SMRs are *modular* because they are small enough to be built in a factory and shipped as a complete unit to site.

Why the interest in SMRs?
Calder Hall in the United Kingdom was the world's first commercial power reactor when it started operating in 1956. With an output of 50 MWe it would be considered as small today. Economy of scale drove the rapid increase in size of power reactors and by the 1970s reactors of 1000 MWe output were common. Today the *average* size of new power reactors under construction is greater than 1,100 MWe and the largest reactors have an output of 1750 MWe. Although they are safe and reliable, there are three problems with large reactors:

- large initial capital cost
- long on-site construction time, typically five years, and
- a 1100 MWe single unit is too large for small grid systems.

SMR Technologies
Many SMRs are based on light water reactors—pressurised water reactor (PWR)/boiling water reactor (BWR) with light water coolant and moderator. These are well-established technologies. Small PWR reactors have been used

for 60 years for nuclear propulsion systems for submarines and icebreakers where reliability is essential. SMRs based on advanced reactor (GenIV) technology are being developed. These have the advantages of operating at higher temperatures enabling higher thermal efficiencies and more process heat applications, including hydrogen production.

Advantages
SMRs have several attractive advantages:

- provide reliable, low emissions power for remote areas or small grid systems;
- because they are small and have less nuclear material in the core, it is easier to design for passive safety;
- the reactor vessel can be installed underground providing protection against external hazards (such as an aircraft crash) and unauthorised access;
- lower initial capital cost and modules can be added while the first modules are operating and generating revenue;
- factory manufactured modules reduce on-site construction time, reducing the risk of project delays;
- when one module is shutdown for refuelling the remaining modules are still producing power;
- SMRs can be designed to load follow to work in a system with variable solar and wind;
- multipurpose—desalination, district heating and process heat in addition to electricity; and
- very compact—the 720 MWe NuScale SMR occupies only 18 hectares.

SMR design based on a PWR
Simply reducing the size of a PWR would result in a higher cost per unit because of the loss of economy of scale. This has to be overcome by factory production and innovative design. Instead of several steam generators separate from the reactor vessel, the steam generators can be wrapped around the core inside the reactor vessel making it an *Integral PWR*. This design also

eliminates the PWR LOCA (loss of coolant accident) fault by eliminating the pipework connecting the steam generators to the reactor vessel.

SMRs based on PWR technology

Russia has a long experience of nuclear powered icebreakers. They decided to install two KLT-40S 35MWe icebreaker reactors on a non-powered barge to provide power for a remote area in Russia. Construction of the 140 metres long and 30 metres wide, 21,000 tonnes displacement barge with its two nuclear reactors was completed in St Petersburg in April 2018. The barge was towed 4000 kilometres through four seas to Murmansk where the fuel was loaded and the reactors tested. In August 2019 the floating nuclear power plant, named *Akademik Lomonosov*, was towed to Pevek, the most northern city in Russia. On 19 December 2019, the floating nuclear power plant started supplying electricity to this remote Artic community and the gold, silver and copper mines in the area. It will also be used to supply district heating.

An example of an SMR at an advanced stage of licensing is the NuScale SMR (United States). The 60 MWe NuScale Power Module (NPM), 25 metre high and 4.6 metre diameter, has a natural circulation primary circuit with integral steam generators. Up to 12 modules can be accommodated in the standard power plant providing 720 MWe(Gross).

The NuScale SMR is in the final stages of Design Certification by the US Nuclear Regulatory Commission (NRC). In 2018, the NRC found that this SMR does not need any emergency electrical supplies because reactor cooling is achieved without them. This is the first power reactor to achieve this. First deployment will be for Utah Associated Municipal Power Systems (UAMPS) Carbon Free Power Project (CFPP) at a site within the Idaho National Laboratory. The NuScale SMR is designed with load following capability, including a 100 per cent turbine bypass for fast load changes.

Advanced reactors

Advanced (Gen IV) reactors provide more opportunities for higher efficiency electricity generation and for process heat. PWRs and BWRs operate at around 300° celsius and are suitable for district heating, desalination and low temperature hydrogen production. Very High Temperature Gas

Reactors (VHTR) operate at up to 1000° celsius which covers a wide range of industrial processes.

Very High Temperature Gas Reactor (VHTR)

The VHTR can be designed to be inherently safe due to its low power density, high heat capacity and unique TRISO fuel. Based on the success of their HTR-10 prototype 10 MW thermal power high temperature gas reactor, which has been operating since 2000, China is now constructing the HTR-PM (Pebblebed Modular) plant at Shandong Shidaowan. Tsinghua University's Institute of Nuclear and New Energy (INET) is the research and development leader for the project. Two 250 MW (thermal power) modules are connected to one 211 MWe steam turbine. The core outlet temperature is a conservative 750° celsius. The fuel is 8.9 per cent enriched HALEU.[1]

Micro reactors

For decentralised generation, microgrids, remote mines, remote communities, military applications and power for critical infrastructure very small reactors with an output of 200 kWe-25 MWe are required. *Micro Reactors*, using heat transfer technology developed for space applications, are being developed.

Westinghouse eVinci micro reactor

Westinghouse is developing the eVinci Micro reactor with Los Alamos National Laboratory (LANL) as a transportable, factory built, combined heat (up to 600° celsius) and power source. The core is a solid monolithic block with three types of channels that accommodate fuel (nominal U Nitride), neutron moderator (nominal Yttrium hydride) and sodium filled sealed heat pipes. Heat pipes were developed by LANL for space missions. They have no moving parts and are designed for long-term operation with no maintenance. They work by a combination of phase transition and thermal conductivity. The sodium evaporates and the pressure difference between the hot vapour pressure and the cold vapour pressure drives the gas to the cold end where it condenses. A wick transfers the liquid by capillary action back to the hot end. Westinghouse is preparing for construction of a eVinci Nuclear Demonstration Unit for installation on site by 2022 and testing in 2023 for commercial deployment by 2025. The United States Department of Defense awarded a contract in March 2020 to Westinghouse to begin design work on a mobile nuclear reactor prototype to support their operations anywhere using standard military transportation.

Molten Salt Reactors (MSR)

One of the Gen IV reactor designs of particular interest to Australia uses molten salt as the coolant. The uranium fuel is dissolved in the salt to form a liquid fuel. There cannot be a reactor meltdown as the fuel is already melted. One of the several companies developing an MSR is Terrestrial Energy (Canada and the United States). The IMSR is designed to produce 600° celsius process heat for industrial use up to five kilometres away, but also for a heat store to provide backup to wind and solar, in addition to electricity generation. IMSR 600° celsius heat for a High Temperature Electrolyser could produce hydrogen at a very competitive cost. The IMSR has successfully completed the phase 1 vendor design review by the Canadian Nuclear Safety Commission (CNSC) and phase 2 is underway. First deployment could be at a site at Idaho National Laboratory (United States) and/or Chalk River (Canada) in the 2020s.

SMRs for Australia

Recent nuclear inquiries in Australia have identified SMRs as being very suitable for Australia's long grid system and remote mines and communities. In 2016, the South Australia Nuclear Fuel Cycle Royal Commission report concluded: 'The smaller capacity of SMRs makes them attractive to integration in smaller electricity markets such as the NEM in South Australia.'[2] In December 2019, the House of Representatives Standing Committee on the Environment and Energy reported on their inquiry into the prerequisites for nuclear energy in Australia[3]. They recommended that 'the Australian Government undertake a body of work to advise on the feasibility and suitability of Gen III+ and Gen IV reactors including small modular reactors in the Australian context'.

Construction of a nuclear power plant is currently prohibited in New South Wales. In June 2019 the Uranium Mining and Nuclear Facilities (Prohibitions) Repeal Bill 2019 was referred to the NSW Legislative Council's Standing Committee on State Development. The committee conducted a thorough inquiry into all aspects of nuclear energy in New South Wales and released its final report in March 2020.[4] The committee concluded that:

The committee could find no compelling justifications from an environmental or human safety point of view which would warrant the blanket exclusion of nuclear energy, especially in its emerging small scale applications, from serious policy consideration in New South Wales.

The report included eight findings and nine recommendations. Recommendation 3 included:

The Department of Planning, Industry and Environment liaise with the Australian Nuclear Science and Technology Organisation to monitor the regulatory approval and commercialisation of Small Modular Reactors in the United States and elsewhere (as appropriate) and report findings to the NSW Government as they become available.

A viable option?

SMRs would provide diversity and reliable electricity generation in Australia. SMRs would make an important contribution to the reduction in greenhouse gas emissions and provide a pathway to reductions in other sectors, particularly process heat for industry and transport. The inherent safety of the latest designs mean they could be located close to the consumer, such as an industrial site. SMRs can load follow to work with renewable energy sources. They would also lead to new industries in Australia.

Endnotes

1   HALEU (High Assay Low Enriched Uranium) meaning enrichment greater than 5 per cent but less than 20 per cent.
2   Final report of the South Australia Nuclear Fuel Cycle Royal Commission https://yoursay.sa.gov.au/pages/nuclear-fuel-cycle-royal-commission-report-release/.
3   *Not without your approval: a way forward for nuclear technology in Australia*, A report of the inquiry into the prerequisites for nuclear energy in Australia, House of Representatives Standing Committee on the Environment and Energy https://www.aph.gov.au/Parliamentary_Business/Committees/House/Environment_and_Energy/Nuclearenergy/Report.
4   Final report of the NSW Legislative Council Standing Committee on State Development inquiry into the Uranium Mining and Nuclear Facilities (Prohibitions) Repeal Bill 2019 https://www.parliament.nsw.gov.au/lcdocs/inquiries/2525/Report%20No%20465-%20Uranium%20Mining%20Bill%20-%20March%202020%20-%20website.pdf.

8 Too cheap to meter or too expensive to matter?

Stephen Wilson

*I*n December 2018, both the CSIRO and the Australian Energy Market Operator (AEMO), referred to commissioned advice from the consulting firm GHD when they published an estimate of the capital cost of small modular reactor (SMR) technology in Australia. The cost was just over $A16,000 per kiloWatt (kW), declining slightly to $A15,823/kW by 2050.[1] This estimate proved controversial—drawing expert criticism from some quarters, and defenders from other quarters—before, during and after recent and continuing inquiries by the Commonwealth, New South Wales and Victorian parliaments.[2] Some wags, re-casting the hackneyed quote from the 1950s that nuclear energy would be 'too cheap to meter', have in recent times taunted their proponents by saying that nuclear power is simply 'too expensive to matter.'[3]

The estimated cost of (large) nuclear reactors in the previous CSIRO report of December 2017 was $9313/kW, declining slightly to $9184/kW by 2050.[4] In January 2018, the chair of the 2006 review taskforce on uranium mining, processing and nuclear energy, Dr Ziggy Switkowski, had been quoted as saying that 'the window [in Australia] for GigaWatt-scale nuclear [reactors] has closed.'[5] Subsequently, awareness of small modular reactor (SMR) designs began to grow in Australia and attention about the prospects for nuclear energy now tends to focus on that emerging class of nuclear technology. The CSIRO responded to the criticism by defending its 2018 estimate and, in December 2019 published an estimate of the unit cost of small modular reactor (SMR) technology in Australia as $A16,304/kW, held constant to 2050 in one scenario, and in the other scenario constant then reducing suddenly to $A7,624/kW in 2032, declining to $A7,145/kW by 2040, then remaining constant to 2050.[6]

The western company most advanced in development of small modular reactors is NuScale Power. Their bottom-up cost estimate, citing work supported by the United States Department of Energy and undertaken by the engineering firm, Fluor, the first major commercial investor in NuScale, is capital cost of $US4,350/kW for the first 720 MW plant with 12 modular reactors, declining to $US3,600/kW for the 'Nth-of-a-kind' plant, excluding warranty, contingency and fees.[7]

What are the real costs? Which engineers' or economists' estimates are more credible? What information do readers need to make up their own minds? How does the capital cost of nuclear power translate into efforts to weigh up the economics of nuclear energy with alternatives for electricity generation? My chapter sheds some light on those important questions and related controversies. It also seeks to go beyond the headline numbers to deepen readers' understanding of what the numbers say and their economic meaning, with a view to making better sense of what can sometimes be heated debate about energy.

Context

Australians now have before them a very large, once-in-a-generation opportunity. The surviving generators from the late 20th century coal fleet in Eastern Australia—a total of 23 GigaWatts (GW) across 16 plants in Victoria, New South Wales and Queensland—are approaching retirement and will need to be replaced, *or refurbished*. The AEMO expects that by 2040, 15 GW (63 per cent of existing capacity) of coal-fired generators will be retired from the system, the distributed energy resources (DER) will increase three-fold, 26 GW of new variable renewable energy (VRE) will be added to the system and somewhere, between 6 and 19 GW of new dispatchable resources, will also be added.[8] Figure 8.1 shows the coal and gas plant retirements, as currently scheduled.

This situation has typically been discussed either as a problem, as an imperative, or as an inevitability. The context is most commonly viewed by media and in public discourse through the lens of climate change. The AEMO

Integrated System Plan makes numerous references to climate policies and climate change effects, but makes no mention of nuclear energy in any of the scenarios it considers. It is not clear whether this is because nuclear energy remains excluded by statutory bans or because of the $A16,000/kW cost number in the annual *GenCost* report—an updated estimate of generation and storage costs produced by the CSIRO and the AEMO.

The multitude of challenges related to climate change—technical, economic and political—are by no means insignificant. But the looming tranche of power plant retirements is, in the first place, age-related. The three-R's problem—retire-and-replace or refurbish—arises from the march of time: it exists quite independently of ambitions and aspirations to reduce carbon dioxide (CO_2) emissions. Australia would have faced the problem in any event. Accordingly, this is best viewed as a national opportunity. Given that this volume is about a nuclear industry future, it is worth also noting that the global fleet of some 450 commercial nuclear reactors is also ageing.

Australia is, therefore, not alone in facing the need—and the opportunity—to think for the long-term, and in a suitably strategic way, about replacement of major power generation assets inherited from latter part of the 20th century.[9] The economics of nuclear energy goes beyond the cost of generation, and cannot be considered in economic isolation. Economic considerations influence and are influenced by technical, commercial, and social, political, legal and environmental considerations.[10]

In Australia, as in many other countries, both the desired and the unintended consequences of energy-related decisions and outcomes will ripple down through the remaining decades of the 21st century. The complex set of concerns, motivations and obligations related to climate change lend themselves to adoption of a long-term perspective that might not otherwise seem so pressing. But there are other reasons that also warrant lifting our eyes to more distant horizons. Not least among these are Australia's national security, the nation's competitiveness, productivity and economic well-being. The level of technological development and the range of Australia's engineering capabilities are important aspects in such long-term thinking. Again, those perspectives both influence and are influenced by economics.

Figure 8.1: Coal- and gas-fired power plant retirements (top) and remaining capacity (bottom)

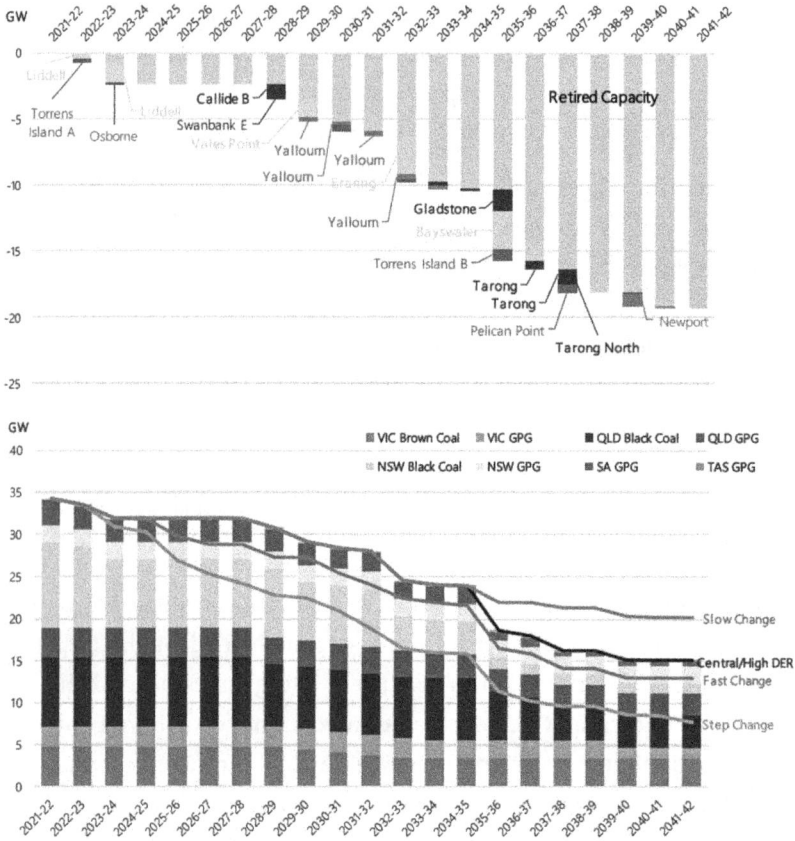

Source: reproduced from AEMO[11]

Replacing ageing coal plants is the lion's share of Australia's task in the electricity generation sector but the opportunity is not strictly limited to that. Less widely appreciated is that not only will most of the existing coal plants have passed their fiftieth anniversary of commissioning between now and 2050, including many during the 2030s. By 2050 almost all of the existing gas-fired plants will also have exceeded their technical service lives (which are shorter than for coal-fired plants), as will all of the existing wind plants, and all of the photovoltaic (PV) panels on rooftops and in utility-scale solar farms. Hydropower plants have far longer technical service lives, and can be assumed to remain in service well past 2050. Hydro replacement is not a concern before 2050. Existing hydro plants are likely to become ever more valuable. The 2020s is a crucial decade in which Australia will either create real options for energy that could be exercised in the 2030s, or pass-up even creating such choices for ourselves.

Such is the immediate energy sector context in which this chapter is written. Following the conclusions is a reflection, looking to the future while drawing in wider contextual factors, related to security, defence, national technological capabilities, and Australia's regional and global responsibilities.

Viable and accurate comparisons

I will compare the economics of nuclear energy with the main alternatives. This is commonly done with the widely-used 'Levelised Cost of Energy' (LCoE) approach. That simple approach is sketched out below in this introductory section, with reference to the controversial cost estimate noted above. Caveats on that particular estimate are briefly described, while noting the limitations, shortcomings, and hazards of the LCoE approach in general. But I need to do more than a simple side-by-side comparison of estimated costs, expressed on a common basis with a metric such as LCoE, calculated from a simple formula.

Just as medicine is not an exact science, nor are engineering and economics. Eastern Australia's National Electricity Market (NEM), like all such competitive electricity systems, are among the most technically and economically complex examples of markets anywhere in existence. Apart from some isolated cases of exceptions that prove the rule—which are usually fairly small systems—no single technology, nor even one category of generation technology can do everything. As a senior Japanese energy official said in 2014: 'we reviewed all energy sources and we found that no energy source is superior in every aspect.'[12]

Almost all electricity grids, therefore, have a *fuel mix*. We should expect that Australia will continue to have a fuel mix for decades to come. These are complex systems composed of many interdependent parts that complement each other, so adopting a system perspective is an important part of our purpose. Our purpose also requires that we have a clear picture of where we are today, how we got here, and where we could go from here, along with the economic and other implications of our choices. The electricity system cannot be discarded and replaced with something else. It needs to perform affordably year-in and year-out, and operate reliably every second of every day.

The prevailing situation

Two decades into the 21st century, Australia has an electricity system with the following characteristics. All five regions in Eastern Australia, being the originally separate state-based systems, are interconnected by transmission as the NEM.[13] Australia has several separate, smaller grids that are each separate 'electrical island' systems, including the South-West Interconnected System (SWIS) and the North-West Interconnected System (NWIS), the systems in Darwin and Alice Springs in the Northern Territory and numerous very small remote area power systems throughout inland Australia.[14] There may be potential for SMR plants in smaller grids and remote locations in future. The focus of my chapter, however, is on the National Electricity Market, warranted by its size, the coal plant retirement opportunity, and the potential to achieve economies of replication with modular plants.

Annual consumption in Australia, including the NEM, changes very little from year-to-year.[15] The NEM system is hydro-thermal with generation-transmission built around large central plants. The now ageing generation plants are largely but not exclusively inherited from the state-owned and centrally-planned systems of the late 20th century. Lignite- and black coal-fired plants located near the mine mouth in Victoria, New South Wales and Queensland form the backbone of the generation fleet, complemented by some combined cycle and open cycle gas plants. Transmission lines from the Snowy hydro plants at the heart of the system strongly link the two large demand centres in Sydney and Melbourne.

The AMEO notes that 'the NEM power system already has 17 gigawatts (GW) of wind and solar nominal capacity installed. Parts of the NEM have among the world's highest levels of wind and solar, including one of the highest levels of residential solar PV.'[16] Of this, 9 GW is rooftop solar, 1.2 GW is utility scale solar farms and 6.8 GW is wind farms. That represents practically all of the new generation capacity added in the past decade. Generation from wind and solar plants has reduced utilisation (the annual capacity factor) of thermal plants, suppressed wholesale prices at various times of the day, increased demands on coal plants to ramp up and down, increasing thermal stresses, accelerating the effects of plant ageing. Those effects, in addition to the more straight-forward effects of lower revenues related to reduced generation, price impacts or both, have resulted in a total of 4,174 MW (all coal-fired, except for 20 MW of gas), being withdrawn from the national market since 2012–13. A further 2,545 MW of capacity withdrawals have been announced, a little over 10 per cent of the approximately 23 000 MW of remaining coal-fired generation capacity in Figure 8.1.[17]

Figure 8.2 shows the installed capacity of coal plants and the total capacity of dispatchable plants overlaid on the evolution of peak, average and base-load demand in the market from 1999 to 2019.[18] The peak varies between about 30 and 35 GW, depending on temperatures. 'S' and 'W' refer to summer or winter peaks, respectively. Contrary to popular myth, the demand never falls below about 15 GW year-round. Even with very high future penetration of behind-the-meter rooftop solar PV generation, base-load demand would be expected to remain owing to demand between sunset and sunrise. The most likely cause of a reduction in the base-load is the retirement of the aluminium smelters in Tasmania, Victoria, New South Wales and Queensland, which account for about 3 GW of the base-load.

Figure 8.2: **Peak, average and baseload demand in the NEM, 1999–2019**

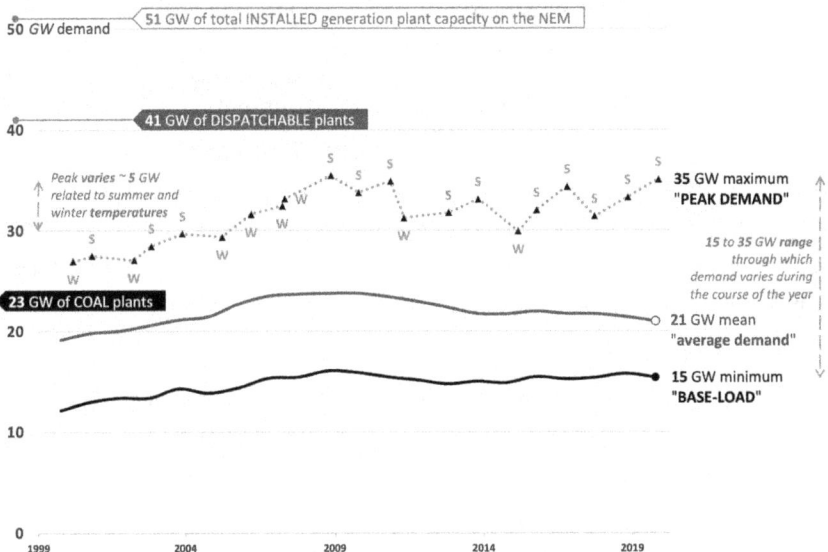

Source: author's chart using data from AEMO[19]

Ownership remains a mix of Commonwealth and state governments and private investors, but the assets comprising the system are largely privatised, with both Australian and international owners from generation, through transmission to distribution networks.[20] Competitive market principles apply to operation of generation assets regardless of ownership. There is a 'wall of separation' between the regulated network businesses and the competitive generation and retail businesses. There are three major commercial players who have vertically integrated their generation and retail supply businesses, providing them with a strong hedge against physical volume risk, and also significantly reducing their exposure to spot and futures market price risk.

The Australian Energy Regulator, a division of the Australian Consumer and Competition Commission, regulates the asset base and allowed rate of return of the network businesses, and oversees the competitive behaviour of generators and retailers, including the large generator-retailers. Under the primary legislation of the National Electricity Law (NEL), the National Gas Law (NGL) and the National Energy Retail Law (NERL), the Australian Energy Market Commission (AEMC) manages the process of accepting or rejecting changes proposed to the National Electricity Rules, the National Gas Rules and the National Energy Retail Rules. The National Electricity Rules are in version 145 as of 21 July 2020, which runs to 1617 pages.[21] The AEMO has the dual responsibilities of system operation and market operation for electricity and gas markets.

In addition to the drivers of the coal plant retirements briefly mentioned above, there are a number of other challenges emerging from the high and rapidly increasing shares of variable renewable energy generation. These include frequency control, system strength, variability and uncertainty (including calls for steeper, larger and more frequent ramping up and down by non-renewable generators, and maintaining security while maximising the potential of wind and solar in the market.[22]

A review of the international literature indicates an extremely wide divergence of estimates of the costs of integrating variable renewable energy—wind and solar—into electricity systems. This is starkly apparent from Figure 8.3. The one characteristic that all researchers, industry experts and practitioners appear to agree on is the general shape of the integration cost curve: costs are low (some sources say negative) at low or very low shares of wind and solar, higher at shares near the capacity factor of those generation sources, and then rise increasingly steeply from the middle of the range and become steepest at high to very high shares.

Figure 8.3: Published estimates of the system cost of integrating wind and solar vary enormously

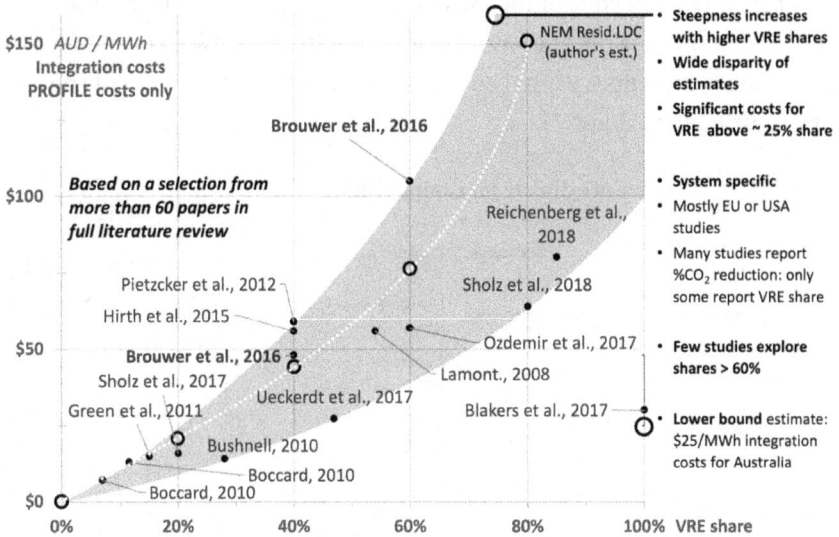

Source: research and literature review by Rioseco (2019), shading and annotation by the author.[23]

Notably, the cost estimates extracted from the literature for Figure 8.3 include only profile costs. Other costs include connection costs to the transmission grid, flow-on effects requiring transmission investments 'deeper' in the grid beyond the direct point of connection, short time horizon balancing costs experienced via regulation, frequency control and ancillary services markets, and in the need for investments or out-of-merit order dispatch of plants to offset declines in system inertia and system strength.[24] Profile costs are related to the need for back-up, energy storage, and the value foregone when available renewable energy cannot be accepted by the system and must be curtailed or spilled; increasing marginal losses as limited transmission capacity that may not be economic to upgrade becomes heavily loaded at times of high renewable energy generation; and payments that must be made to compensate customers for interruptions or other actions known as 'demand response.'

Estimates and headline numbers in context

I began this chapter with the controversy on estimates of the capital cost of deploying nuclear energy in Australia. Readers may wonder why the views of engineers, scientists and economists diverge on the estimated capital cost of building SMRs in Australia? This is a good question. The up-front construction cost is certainly one of the most important numbers in the economics of nuclear energy. It is not the whole story but it is a good place to start. When an engineer presents a capital cost estimate, it is good to ask about the class of the estimate (Table 8.1).

Table 8.1:   Classes of estimate for capital costs

Class	Level of Definition (% of complete def.)	End Usage	Expected Accuracy	Contingencies Project	Process (i)
5	0% to 2%	Concept screening	Upper: +30% to +100% Lower: -20% to -50%	30%	New Concept with Limited Data: 40% to 100%
4	1% to 15%	Study or feasibility	Upper: +20% to +50% Lower: -15% to -30%	25%	Concept with Bench Scaled Data: 30% to 70%
3	10% to 40%	Budget authorisation or control	Upper: +10% to +30% Lower: -10% to -20%	20%	Small Pilot Plant Data: 20% to 35%
2	30% to 75%	Control or bid/tender	Upper: +5% to +20% Lower: -5% to -15%	15%	Full Size Models Operated 5% to 20%
1	60% to 100%	Check estimate or bid/tender	Upper: +3% to +15% Lower: -3% to -10%	10%	Process Used Commercially 0% to 10%

Source: Association for the Advancement of Cost Engineering[25]
Note:

(i)   The state of technology development may be independent of the level of project definition. So, for example, a completely defined project (Class 1 estimate) but using a new concept with limited data might have an expected accuracy of (say) -5% to +10% and a project contingency of 10%, but a process contingency somewhere between 40% and 100%. Conversely, a concept screening with 2% level of project definition but using a process used commercially would have an expected accuracy of (say) -35% to +50%, and a project contingency of 30% but a process contingency of (say) 5%.

There are a number of other good questions.

What is this estimate intended to be used for?

Is the estimate intended to be conservative, neutral or aggressive?

Is the estimate intended to be a 'P50' central estimate, with a 50–50 likelihood of being whether high or low, a conservative P80 estimate, with only a 20 per cent chance of being too low, or an aggressive P20 estimate, with only a 20 per cent probability of being too low?

Has an uncertainty range been provided around the estimate?[26]

Is the range appropriate for the class of estimate?

What are the inclusions and exclusions for the estimate?

What is the maturity basis or number in the series: first-of-a-kind versus nth-of-a-kind, globally (relevant to the process or technology maturity) and locally (relevant to relevant project-specific experience)?

What will be the cost estimators' personal gain (or loss) if they are right or wrong? (Is there any 'skin in the game'?)

Who paid the estimator to make the estimate?

Did the sponsor or the estimator want one or other technology or project to succeed or fail or to appear attractive or unattractive - and why?

Does the cost estimator believe the technology or project will succeed or fail—and why?

The next three sections are largely interpretative. The first stretches the canvas. I will go beyond the headline numbers of the simple economics sketched out in the introduction, setting out the scope, methods of analysis, and sources of data. Some material to assist readers without a background in economics or engineering or the electricity industry is presented alongside the main thread of the section: on the terms used in the analysis; on the importance of time horizons, of scale, and constraints in the real world; on the development of technology, and lessons from megaprojects; some analytical frameworks are described, analysis is used to make sense of numbers, and in recognition that engineering and economics are not exact sciences, the role of judgment is described.

The second section is an analysis of the data and analysis presented in going beyond the headline numbers. Economics alone can be dry, theoretical, unsatisfying and ultimately impractical. To be of use to the public, whether as investors, as consumers, as employees, as voters, or simply in the most important role as interested members of Australian society with a shared stake in the future, the public need to go beyond economics-on-paper. It is necessary to ground the public conversation in real world questions, considering ownership control of assets and investment financing, and seeking consensus on a preferred balance of planning and co-ordination, of competitive forces and regulation. Those real-world dimensions feed directly back into high level policy questions on market design and industry structure.

The third section contains my conclusions and some reflections that offer a perspective on the analysis and exploration of the economics of nuclear power compared with alternatives for electricity generation. The opportunities presently before Australia are set out below. They include viable options for the current generation and the next one or two at least, Australia's current capabilities as a nation, and how the country might enhance them. The economics of nuclear energy and the alternatives are crystalised as scenarios describing contrasting visions of the future.

Going beyond the headline numbers

This section is written with inquiring readers in mind: non-specialists interested in the numbers. Those with a background in power system engineering or economics may find some parts straight-forward and the subsequent analysis and conclusions of greater interest. I encourage readers to compare estimates of generation costs with prices familiar to them, while remembering that the *estimates* of *future* generation *costs* in this chapter are one factor among many in the *wholesale* market, while the *actual historical or current* cents per kiloWatt-hour (¢/kWh) number on their home electricity bill is a *retail price*, inclusive of generation, network, retail and other costs, as well as any additional profit margins (or losses) along the value chain and any taxes or subsidies (explicit or implicit). The Australian Consumer and Competition Commission (ACCC) has published helpful charts that break average residential electricity bills and energy prices down into their component parts and show how they have changed over a decade.[27] According to the ACCC's analysis, the wholesale energy component of average residential prices increased from 7.3¢/kWh to 10.1¢/kWh between 2007–08 and 2017–18, while total average residential prices increased from 19.0 to 29.6¢/kWh during the same period.

Wholesale market prices in the National Electricity Market are quoted in dollars per MegaWatt-hour ($/MWh), as are generation cost estimates. To convert to ¢/kWh, simply divide by 10. For example, $100/MWh is the same as 10 ¢/kWh. The cost of serving a residential electricity consumption pattern is higher than business customers, so the wholesale energy component of residential bills is higher than the average price in the wholesale market. Energy generation is one component among five in delivered prices, all of which increased between 2007–08 and 2017–18. The volume-weighted average price in the wholesale spot market in 2007–08 was $54/MWh (regional averages ranged from a low of $A45 in the New South Wales market to a high of $A101 in South Australia) and increased to $87/MWh in 2017–18 ($A75 in Queensland and $A111 in South Australia).

Capital costs and costs per unit of electricity

I have mentioned that LCoE is a common method used to compare one type of generation with another. This method was developed to enable comparisons to be made on a common basis between, for example, a hydro power plant with very high up-front capital costs and very low standing operating costs and a coal plant with much lower initial fixed costs and much higher variable costs for fuel, operation and maintenance. The same method might be used to compare the costs of a nuclear power plant with a gas-fired power plant or, indeed, with a hydro or coal-fired plant. LCoE is recommended for simple ranking when on a limited budget, is suitable for high-level screening (accept/reject decisions), but is not recommended for selecting from mutually exclusive alternatives.[28]

Even in cases where it is recommended or acceptable, due care is needed in its application. While the method allows for technologies with dissimilar capital and operating costs to be compared on a common basis, other characteristics also need to be comparable, including: dispatchability, annual capacity factor, and the technical service life, transmission connection costs, and contribution to (or effect on) the system as a whole, as well as emissions or other externalities. If those characteristics are not the same or very similar, the comparison implied by the LCoE numbers will be misleading. CSIRO (2019) notes that:

> The levelised cost of electricity (LCOE) is a simple screening tool for quickly determining the relative competitiveness of electricity generation technologies. It is not a substitute for detailed project cashflow analysis or electricity system modelling which are both better suited

to representing real electricity generation project operational costs and performance.

'Relative competitiveness' is probably an over-statement: there are a multitude of factors, many subtle, that influence the competitiveness of a power plant in any given system and market. The 'approximate relative economic attractiveness' would be more appropriate. Competitiveness and commercial and financial attractiveness need to be considered separately.

Table 8.2a provides an example of a calculation with approximations and simplifications, using inputs and assumptions from CSIRO (2019), to show how one can arrive at an LCoE number of the kind in that report. The value of $A332 is very close to the 'high' value of $A333/MWh in CSIRO (2019).

Table 8.2a: **LCoE worked example with CSIRO GenCost (2019) inputs (high cost case)**

Reactor module capacity, thermal		133	MWth
Thermal efficiency		45%	
Reactor module capacity, electrical-gross		60	MWe
Nº modules and plant capacity	12	720	MWe
Plant capacity on a sent-out basis		683	MWe
Series Nº & CapEx on a per unit capacity basis	*Nth*	16.000	AU$M /MW
CapEx on a total plant overnight cost basis		11,520	AU$M
Construction period	5 years	60	months
Interest During Construction (IDC) approximate		1,728	AU$M
CapEx including capitalised IDC @WACC		13,248	AU$M
Project Contingency	30%	3974	AU$M
Process Contingency (technology maturity)	10%	1325	AU$M
CapEx including IDC and Contingencies		18,547	AU$M
Technical service life		60	years
Weighted Average Cost of Capital		6.00%	
Capital recovery period		30	years
Fixed operation & maintenance per unit of capacity		$200,000	/MW
Capacity charge as an annuity		$1,871,436	/MW /y
Plant Capacity Factor		80%	of capacity x 24h/d x 365d/y
Annual generation sent out		6,648	MWh /y per MWe of gross generating capacity
Capital recovery charge		$281.51	/MWh
Annual fixed O&M		$30.08	/MWh
Fuel + variable operation & maintenance		$20.00	/MWh
Long-run marginal cost over capital recovery period		$331.59	/MWh

Source: author's calculations using inputs and assumptions from CSIRO (2019)
Note: values in boxes are inputs.

(i)   The source of this thermal efficiency is not clear.

(ii)   A discount rate or weighted average cost of capital (WACC) is not mentioned in the CSIRO report.

Table 8.2b changes several input parameters, as follows:

- the thermal efficiency is corrected from 45 per cent to 30 per cent;[29]
- unit capital cost is reduced from first-of-a-Kind $A16,000/kW to nth-of-a-kind $A4,800/kW;
- a 12-module plant 'overnight' capital cost falls from $11.5 billion to below $3.5 billion;
- learning and modularisation reduce construction time from 60 months to 36 months;
- the weighted average cost of capital is reduced from 6.0 per cent annually to 5.3 per cent (real);

Table 8.2b: LCoE worked example with inputs for illustration (low cost case)

Reactor module capacity, thermal		200	MWth
Thermal efficiency		30%	
Reactor module capacity, electrical-gross		60	MWe
Nº modules and plant capacity	12	720	MWe
Plant capacity on a sent-out basis		683	MWe
Series Nº & CapEx on a per unit capacity basis	*Nth*	4.800	AU$M /MW
CapEx on a total plant overnight cost basis		3,456	AU$M
Construction period	3 years	36	months
Interest During Construction (IDC) approximate		275	AU$M
CapEx including capitalised IDC @WACC		3,731	AU$M
Project Contingency	10%	373	AU$M
Process Contingency (technology maturity)	0%	0	AU$M
CapEx including IDC and Contingencies		4,104	AU$M
Technical service life		60	years
Weighted Average Cost of Capital		5.30%	
Capital recovery period		60	years
Fixed operation & maintenance per unit of capacity		$100,000	/MW
Capacity charge as an annuity		$316,359	/MW /y
Plant Capacity Factor		90%	of capacity x 24h/d x 365d/y
Annual generation sent out		7,479	MWh /y per MWe of gross generating capacity
Capital recovery charge		$42.30	/MWh
Annual fixed O&M		$13.37	/MWh
Fuel + variable operation & maintenance		$10.00	/MWh
Long-run marginal cost over capital recovery period		$65.67	/MWh

Source: author's calculations using inputs and assumptions from CSIRO (2019)
Notes:
(i) NuScale (2019) report the thermal output of their design as 200 MW$_{th}$, giving an implied thermal efficiency from heat to power of 30%. The 'industry standard' thermal efficiency assumption applied to nuclear power in the statistics of the International Energy Agency is 33 per cent.
(ii) This is based on NuScale's published estimate of $US3,600/kW for an Nth-of-a-Kind plant, converted at 0.75AUD/USD. This is for illustrative purposes. The correct way to convert costs from an American site to an Australian site is far more complex and beyond the scope of this chapter.

- those changes reduce interest during construction from $1.7 billion to $275 million;

- the capital cost including IDC is reduced from $13.2 billion to $3.7 billion;

- project contingency of 30 per cent and process contingency of 10 per cent (new technology or design) reduce contingency from $5.3 billion to $373 million;

- total capital cost reduces from $18.5 billion to $3.7 billion;

- the capital recovery period is extended from 30 years to match the 60-year service life;

- the plant capacity factor is increased from 80 to 90 per cent, increasing annual generation;

- fixed operation and maintenance costs are reduced from $200/kW to $100/kW; and

- fuel and variable operating and maintenance costs are reduced from $20 to 10/MWh.[30]

The $A16,000/kW capital cost has been interpreted here as a first-of-a-kind cost. Capitalisation of interest during the five years of construction alone adds $A1.7 billion to the project, and 40 per cent in project and process contingencies add a further $A5.3 billion, for a total of $A7 billion on top of the overnight cost of $A11.5 billion implied by applying the unitised capital cost to the capacity.

The combined effect of those changes is to reduce the LCoE of a 12-module SMR plant from over $A330/MWh to about $A65/MWh. The point of these two worked examples is not to say that version 2b is correct and version 2a is wrong. It is to show how easy it is to come up with very different results—one above the cost of small diesel peaking plant generation and very uncompetitive, the other below the cost of new gas-fired plants and highly competitive—and to show how sensitive the results are to changes in the inputs. What matters in the end is not academic debates about these numbers. What matters is the results that actual engineers and project developers are able to achieve on real projects. The insight from the worked example is that nuclear energy could be competitive in Australia.

Twenty years ago, the costs of wind power and solar power were nowhere near where they are today. A similar prospect exists today for nuclear energy from small modular reactors to be able to generate near-zero emissions electricity at competitive cost levels with high reliability.

A wider context of consideration

Although my focus is the electricity grid, a major part of the wider energy ecosystem, electricity is far from the whole energy story.[31] Each power plant in the electricity grid is one part of a complex, interconnected system, which must operate at precisely the same frequency of 50 cycles per second *and in full synchronisation*, otherwise the entire system is subject to collapse in less than a minute: a cascading blackout. To achieve this settled state, 'electrical current ... must be produced, to the millisecond, at the moment of consumption, giving an exact balance between power supply and demand. Stable power grids are based on this principle'.[32] This inescapable fact is foundational to the economics of electricity, and its implications are essential to appreciate in any discussion of the economics of nuclear energy and alternatives for electricity generation.

Electricity grids need to balance supply with demand across time scales that span twelve orders of magnitude: from fractions of a second to decades. Scale is significant in energy systems, and energy system scale also spans twelve orders of magnitude. Some examples to help visualise time and energy across twelve orders of magnitude are provided in Box 8.1. Box 8.2 describes units used throughout this chapter and in references, including the relationship between power and energy.

A seller's prices are his buyer's costs, and so on down the value chain. The interchangeable use of the word *cost* and the word *price* can give rise to confusion, risking misunderstanding and conversations and debates ending up at cross-purposes. In this chapter, the term *cost* is used for the capital and operating costs incurred by the owner of a power generation asset, and for other costs borne by other entities in the supply chain upstream of the customer. Unless otherwise stated (for example, when referring to 'retail prices'), the term *prices* is used for the spot prices paid by the demand side to generators in the wholesale electricity market.

In considering the scope for deploying nuclear energy in Australia, the opportunities flowing from deployment of small modular reactors (SMRs) require closer attention. Also essential is understanding the 'ecosystem' into which nuclear energy or any other technology would be deployed.[33]

Methods of visualisation and analysis

There is much behind the claim: 'Wind and solar plants will soon be cheaper than coal in all big markets around world.'[34] Generation technologies differ widely, not only in the levels of their respective costs, but also in the *structure* of their costs. Some technologies have low up-front costs and high variable costs for fuel, operation and maintenance. Small diesel generators (similar to large truck engines) and open-cycle gas turbines (similar to aircraft engines) are an example: cheap to buy and expensive to run. Other technologies are the reverse: expensive to build but cheap to run. Further complicating comparisons, some technologies have short service lives of perhaps 10 years for battery storage, 20 years for solar panels and 25 or 30 years for gas turbines; while others have long lives of 40 to 50 years for coal-fired plants and 40 to 60 or more years for nuclear power plants.

Finally, one technology may find a different role from another technology in a power system. The annual capacity factor or average utilisation is one indicator of this. There may be partial overlap between the capacity factor of two different plants. The validity of any comparison must be rigorously tested to both meaningful and unbiased.

Figure 8.4 in panel a) shows different cost structures graphically; panel b) shows how costs map onto a load-duration curve; panel c) shows how the situation changes when non-dispatchable variable renewable energy takes priority in dispatch 'eats into' the load-duration curve. Dispatchable plants are then left with the task of balancing the 'residual load-duration curve: not only varying customer demand, but its combination with variable renewable energy.[35]

I have noted that in competitive electricity markets, a price-duration curve can be created. It has the same general shape as the load-duration curve but is far more extreme. In the case of Australia's National Electricity Market, the price is allowed to go up to a spot price cap of $A15,000/MWh, and down to a floor of *negative* $A1,000/MWh. Those extreme prices far exceed the

Figure 8.4:   the effect of variable renewable energy on the load duration curve

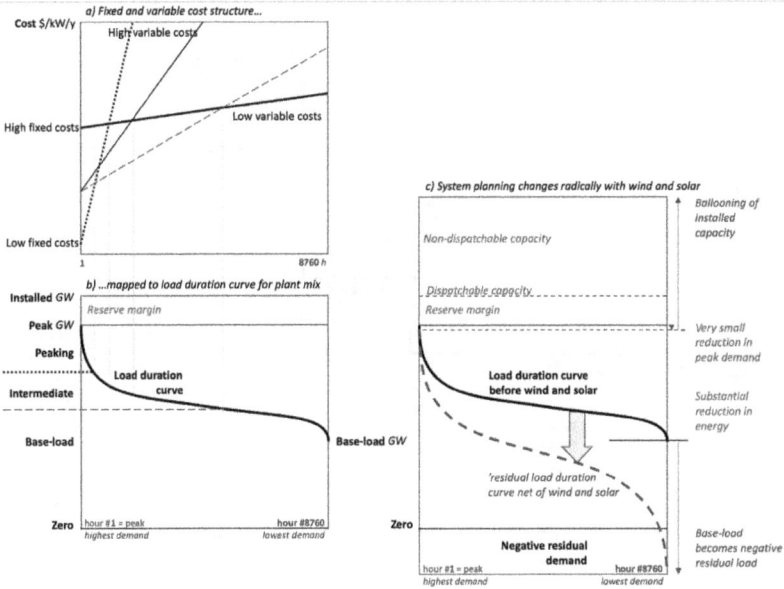

Source: my diagrams were prepared with reference to Australian and other experiences.

underlying costs. When such prices occur, they reflect either scarcity, or abundance, respectively. The economic philosophy of this market design is that high price events can signal to investors to invest in new peaking plant (or energy storage), or to offer demand-side responses (turning off, reducing, or 'interrupting' loads in return for a payment).

The underlying cost structure of the system can be visualised as a short-run marginal cost curve. A simplified version of this is presented in Figure 8.5. This assumes that plants of a similar type all have the same fuel and variable operating and maintenance costs. In reality, each plants' costs are a curve, reflecting part-load thermal efficiency curves and numerous other more complex factors.

Three cost curves are shown: with only thermal plants, with hydro and thermal (dispatchable) plants, and with all capacity available. In the base-load to average demand level, if all generators offer at their short-run marginal cost, the price outcome will be set by Queensland or New South Wales black coal (about $35 /MWh). If large quantities of wind and solar are able to generate,

Figure 8.5: the structure of demand and short-run marginal cost of generation

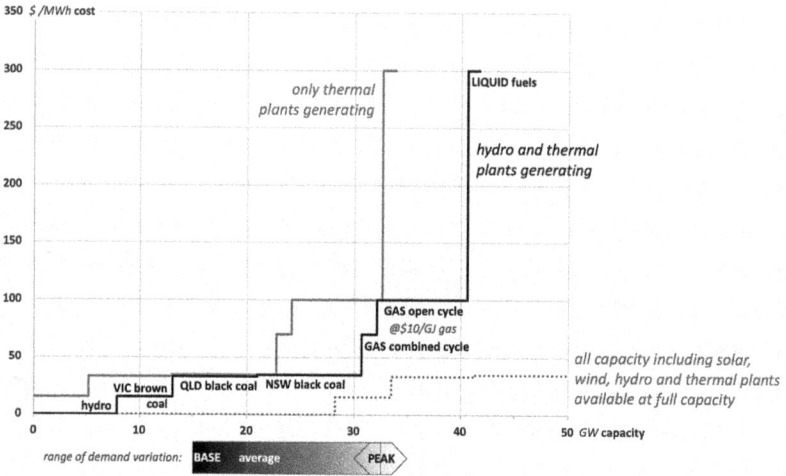

Source: my diagrams were prepared with reference to approximate Australian cost levels.

Figure 8.6: Mainland National Electricity Market–average offers July 2015, July 2017, March 2018

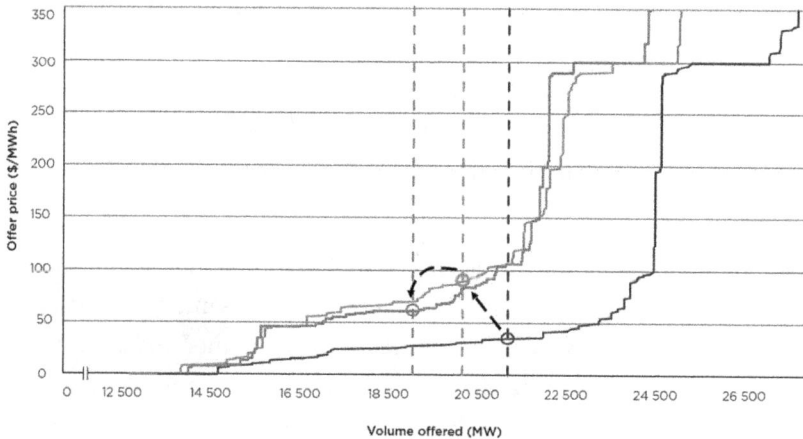

Source: ACCC analysis of AEMO data.[36]

the price will be pushed to zero, or potentially negative (generators must pay the market for the privilege of generating, off-takers such as retailers, receive payments for load, rather than paying for load). Generators able to earn revenue from Renewable Energy Certificates must be generating to

Figure 8.7: **superimposing offer curves on the illustrative short-run marginal cost curves**

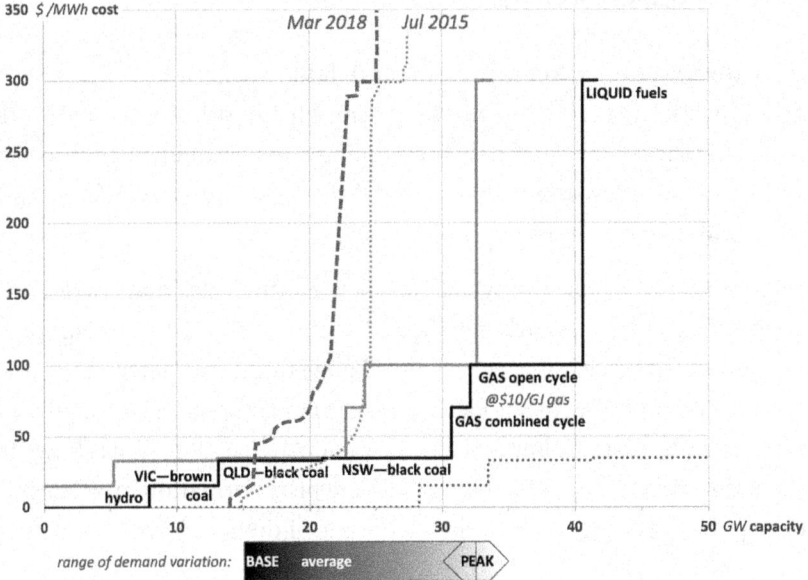

Source: author's approximation of the ACCC curves superimposed on the author's cost curve chart.

create the certificates, so will be willing to offer the market negative prices to get some revenue. Thermal generators (such as coal and combined-cycle gas plants) whose owners would prefer to avoid the costs of shutting down and restarting may offer negative prices for the capacity up to the minimum stable generation level of each unit they wish be generating.

Around the peak demand level of 30 to 35 MW, the short-run marginal cost may be anywhere between about \$A15/MWh illustrative of Victorian brown coal, \$A100/MWh (from an open cycle gas turbine paying \$A10/GJ for its gas, to \$A300/MWh for a diesel engine or turbine on liquid fuels. Mental arithmetic can be applied for the effect of lower gas prices (as at present).

Generators indeed construct their offer curves along the lines described above, as shown in Figure 8.6. Figure 8.7 superimposes on Figure 8.5 the offer curves in Figure 8.6. Not only do generators offer at or below zero price out to the region around the level of demand, but the rising portion of the offer curves are to the left of the cost curve. That shows a market that tightened after retirement of Hazelwood, and perhaps a degree of market pricing power for generators.

Visual analysis of market price-demand data

Price-demand data from the market represent the price outcomes in the wholesale market for each half-hour, being the last offer in each period required to meet demand in that period. Prices are where the offer (supply) and load (demand) curves intersect.

Figure 8.8 is the half-hourly data for New South Wales that corresponds to two of the national market mainland region offer curves in Figure 8.6. The uplift in the price level, as well as a change in the character of the price-demand outcomes is evident. Panels a) and b) of Figure 8.9 show equivalent data for South Australia for August 2009 and August 2019. This shows the radical transformation of how the wholesale price formation mechanism is working that has occurred during the decade of the build-out of renewable energy capacity.

The same pattern is evident in all other months of the year, and already affects all states of the NEM, although to a slightly less dramatic extent than in South Australia. Panels c) and d) of Figure 8.9 show how the price spreads across the South Australia-Victoria interconnector have also changed over the same decade. Prices that were previously very closely correlated now exhibit very large excursions from the line of arbitrage. This figure indicates that interconnector capacity limits are reached, or other constraints (such as system strength requirements in South Australia) prevent traders from arbitraging the price differences.

The opportunity

The potential contribution of nuclear power to the market is clear and straightforward: ahead of old coal plants retiring (Figure 8.1), civil works are prepared, leverage sites close to existing transmission connections. As coal units are retired and the capacity-demand balance requires, reactor modules are added to the common plant infrastructure.

The nuclear plants 'slot in' to Figure 8.5 at a short-run marginal cost of about $10/MWh, just under Victorian brown coal plants and well below NSW and Queensland black coal plants. To be able to recover their capital

Figure 8.8   Spot prices, and demand in New South Wales, July 2015 and July 2017

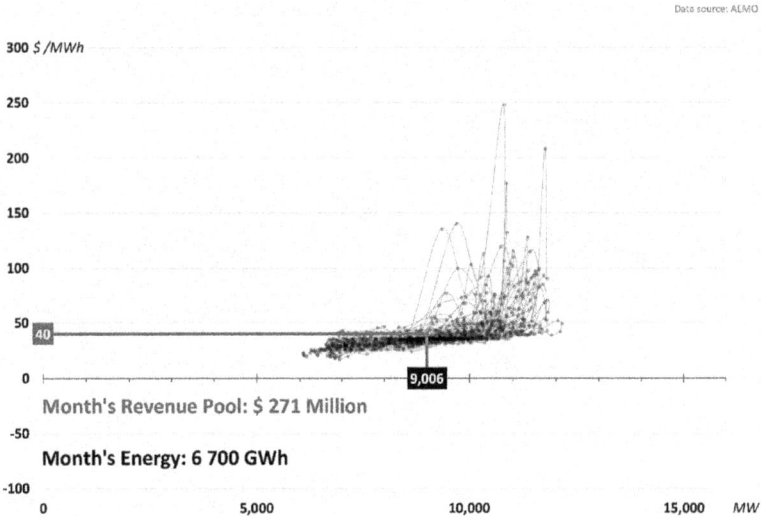

Data source: AEMO

Month's Revenue Pool: $ 271 Million

Month's Energy: 6 700 GWh

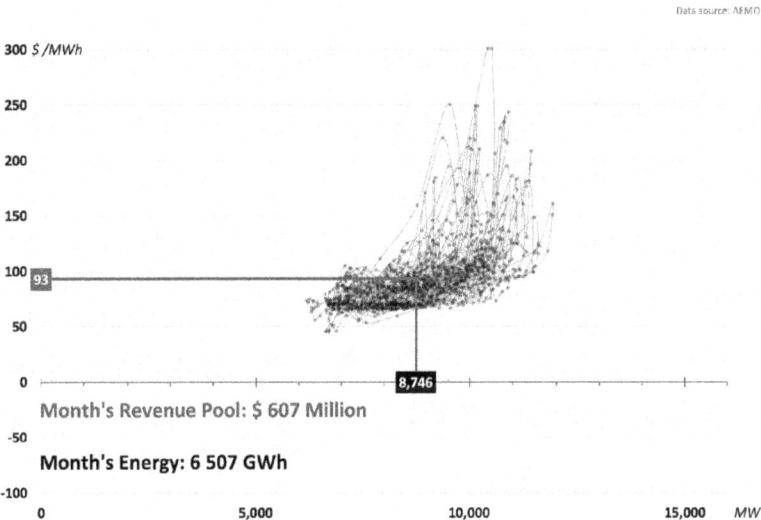

Data source: AEMO

Month's Revenue Pool: $ 607 Million

Month's Energy: 6 507 GWh

Source: author's charts from AEMO price and demand half-hourly data[37]

cost in the electricity market, however, they will require a price of at least $A65/MWh (Table 8.2 b) and probably closer to $A100/MWh. That implies that open cycle gas power plants will need to be setting the price for most of the time, with a gas price of around $10/GJ. In this sense, nuclear power

Figure 8.9: **Spot prices and demand in South Australia and prices across the interconnector**

August 2009

a) half-hourly wholesale prices versus demand

August 2019

b) half-hourly wholesale prices versus demand

c) half-hourly prices in SA versus Victoria

d) half-hourly prices in SA versus Victoria

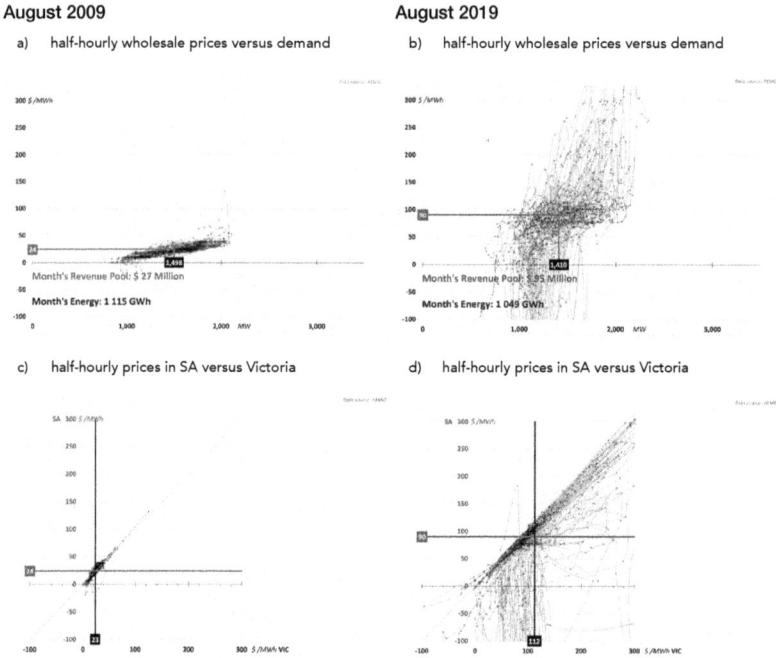

Source: author's charts from AEMO price and demand half-hourly data[37]

can be seen as a hedge against gas prices (as it is in the United Kingdom, for example), or as a lever to maximise income from future exploration, production and development of gas as LNG for export to Asia. It is also as a technology that nuclear power can make a major contribution towards the long-term aspiration of achieving net zero emissions, while maintaining high levels of reliability, system strength and all of the other issues that are rapidly becoming major challenges with the rapid increase in the deployment of wind and solar power.

The investment challenge

There is, as might be expected, a large market and financial wrinkle that is a threat to realisation of the very rational and sensible engineering and economic opportunity. It is captured in a quote from a banker with many decades of experience financing the Australian electricity sector: 'no-one can produce a bankable price forecast of the Australian electricity market today'. This view is held by all major Australian and international banks active in financing Australian infrastructure, energy and resources projects.[38] Project financing requires a power purchase agreement with an investment-grade offtaker, a sufficiently long tenure, and able to ensure the capital recovery of the entire plant capacity. Wind and solar projects are the only types of investment that have been undertaken for the past decade. They have been the only kind of projects that are bankable. That is not only because of carbon policy risk (although that is a factor) and it is not only because those projects can secure a 'revenue stream from certificates outside the electricity market.' It is also because they have been able to contract with a large creditworthy off-taker, typically one of the generator-retailers with a renewable energy certificate liability, access capital from ARENA and the CEFC, and because the financial structure enables them to recover their capital within a seven-year loan tenor, or certainly within two seven-year loans (with the first including the construction period).

Nuclear power will require far longer financing periods, probably for a minimum of 30 years. To secure financing, nuclear power plants, including small modular reactors, will also require some form of secure, bankable long-term power purchase agreement guaranteeing a stream of revenue large enough to recover the capital investment at the cost of capital. This problem was foreseen as early as 1983:

Although it might be possible to raise capital on reasonable terms to build base-load generating plants that would not be insulated by long-term contracts from the natural risks of the bulk power marketplace, we find it hard to imagine that base-load power plants anything like those we see today would be constructed in the face of the extraordinary additional opportunity risks inherent in a regime permitting only spot market sales.[39]

This problem relates to the structure and functioning of the market, and its consequences are more far-reaching than academic debates about the LCoE of one technology or another.

Part of the spirit associated with the electricity market reform movement in Australia during the 1990s was the idea that governments could pass a well-written piece of electricity legislation, design a carefully thought out market expressed in a set of market rules, establish a virtual 'wall of separation' between the 'contestable' (energy) and 'natural monopoly' (network) businesses, set up appropriate independent regulators, then step back, and never have to worry about electricity again.

Such an approach is no longer viable. Government reinserted itself through renewable energy targets.[40] At the first signs of trouble in the market, governments are in response mode. One blackout is sufficient to dispel any thoughts of 'leaving it to the market.'

A 2050 strategy

Retirement of existing thermal (mainly coal-fired) plants creates an opportunity for between 10 and 20 GW of small modular reactors to be deployed in Australia. The first step in any series of nuclear deployment goals needs to be the repeal of the legislative bans on nuclear energy in Australia. New, dispatchable, low emission generation will be valuable as each successive plant in the ageing coal fleet nears and passes its retirement date. Physical deployment of nuclear generation in the 2030s will require substantial development and preparatory work in the 2020s by governments, universities, developers, investors, the public, consumers, banks and regulatory agencies. Natural interim goals centre around commencing those tasks now to create real options for deployment.

Options and choices

The trends since the early nuclear power reactors entered service in the 1950s are set out in Figure 8.10:

- pursuit of economies of scale, from 60 MWe scale to 1000 or more MWe;

- boiling water reactors (BWRs) and pressurised water reactors (PWRs); and

- eventual dominance of PWRs (more than 4-to-1 in operable capacity).

Economies of scale have proved elusive for a number of reasons. Economies of scale via large reactors are now being complemented by pursuing economies of replication, fabrication over construction, plant modularisation and emergency planning zone downscaling through small modular reactors. The lower end of the SMR range is 60 MW_e: the same capacity as the first commercial

Figure 8.10: **Global nuclear energy plants by type and status**

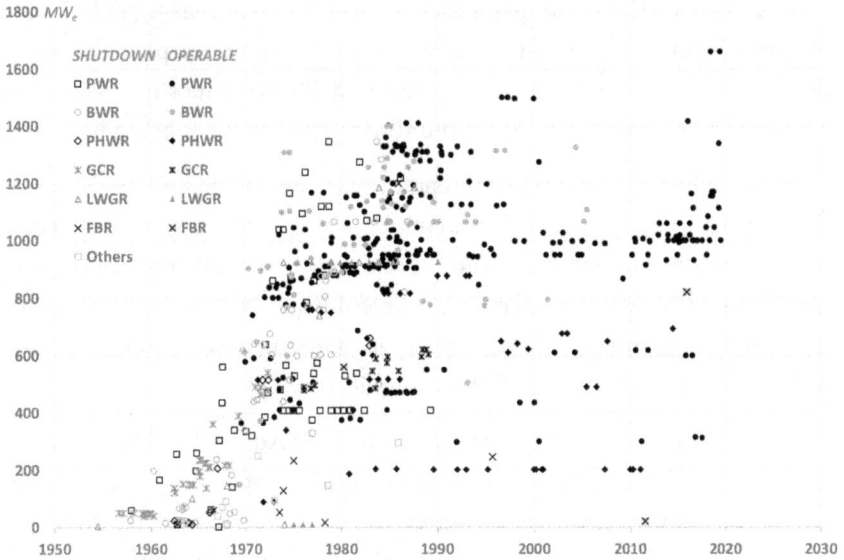

Source: author's chart from World Nuclear Association reactor database[41]

demonstration reactor in the United States. The design and safety of the new reactors incorporates 70 years of engineering development and lessons from design, planning, operation licensing, and approvals.

In 2002, the International Energy Agency based in Paris (not to be confused with the IAEA located in Vienna), citing the United States Department of Energy (2002) as the source, included a page on 'Advanced Nuclear Reactors' with a table of six 'Generation IV Systems and Best Deployment Date, which ranged from 2015 for the Sodium-Cooled Fast Reactor to 2020 for the Very High Temperature Reactor and 2025 for each of the other types: Gas-Cooled Fast Reactor, Molten Salt Reactor, Supercritical-Water-Cooled Reactor and the Lead-Cooled Fast Reactor. Nearly twenty years later, it is a smaller, simpler variant of the Pressurised Water Reactor, the NuScale SMR that has emerged as the closest to deployment. The project was supported by the United States Department of Energy since the late 1990s. Perhaps not as technically intriguing as the advanced reactor designs, it was not mentioned at all in the 2003 edition of the *World Energy Outlook*.[42]

An academic reactor or reactor plant almost always has the following characteristics: (1) It is simple. (2) It is small. (3) It is cheap. (4) It is light. (5) It can be built very quickly. (6) It is very flexible in purpose ('omnibus reactor'). (7) Very little development is required. It will use mostly 'off-the-shelf' components. (8) The reactor is in the study phase. It is not being built now.

On the other hand, a practical reactor plant can be distinguished by the following characteristics: (1) It is being built now. (2) It is behind schedule. (3) It is requiring an immense amount of development on apparently trivial items. Corrosion, in particular, is a problem. (4) It is very expensive. (5) It takes a long time to build because of the engineering development problems. (6) It is large. (7) It is heavy. (8) It is complicated.[43]

This is an extract from a short 1953 memo by Admiral Hyman Rickover who directed the first commercial demonstration of a PWR in the United States. The NuScale SMR is also a pressurised water reactor: the dominant reactor type in the world and, importantly, each module is the same scale as the first American PWR.

Table 8.3 shows three scenarios for such deployment, with numbers of plants, modules, and capacities by state.

Table 8.3   Examples of small modular reactor deployment potential in to replace retiring plants

	Low case				Mid case				High case			
	Plants #	Mod 12	Capacity MW	GW	Plants #	Mod 12	Capacity MW	GW	Plants #	Mod 12	Capacity MW	GW
Qld	5	60	3 415	3	6	72	4 098	4	9	108	6 147	6
NSW	6	72	4 098	4	8	96	5 464	5	12	144	8 196	8
Vic	3	36	2 049	2	4	48	2 732	3	6	72	4 098	4
SA	1	12	683	1	2	24	1 366	1	3	36	2 049	2
	15	180	10 245	**10**	20	240	13 660	**14**	30	360	20 490	**20**
WA	1	12	683	1	2	24	1 366	1	1	12	683	1
NT									1	12	683	1
AUS	16	192	10 928	11	22	264	15 026	15	32	384	21 856	22

Source: my examples illustrating deployment scenarios.

Construction of between 10 and 20 GW of nuclear capacity between 2030 and the 2050s would involve deployment in the long-term of 15 to 30 plants each with up to 12 60 MW reactor modules, or of a smaller number of plants, comprised of larger SMR modules of up to 300 MW each.

Some conclusions

Australia is at an important juncture for its energy sector, for the electricity sector in particular, and by extension for the economy as a whole. The opportunity is before us. The system is facing threats and challenges, on costs and reliability, that are arguably becoming more challenging. Emissions could be added to the list, particular if deep reductions are to be achieved in a long-term time frame. Serious consideration is warranted.

It is not difficult to play spreadsheet games or run modelling exercises to suggest that nuclear energy makes no economic sense. As this chapter has shown, it is just as easy to play similar games to suggest that nuclear energy is the most appealing technology for long-term deployment into the backbone of the electricity grid, and potentially for isolated grids and remote sites as well. That would be to do the nation a disservice. Debates about 'energy policy' in Australia have really been a disguise for disagreements about energy doctrine. When people say that the government does not have an energy policy, what that really boils down to is that we cannot reach agreement on an energy doctrine. Defence people talk about military doctrine. Diplomats and academics talk about foreign policy doctrine.[44] Once an energy doctrine is settled, energy policy will readily follow.

Capabilities

Australia has many natural and societal advantages and capabilities that would be valuable for the deployment of nuclear energy. They include: a high income and well-educated population; a small cohort of world class nuclear engineers and scientists; university courses in nuclear science and engineering; a tertiary education system with a number of institutions in the top one per cent of universities globally; engineering firms and engineers with experience in delivering large projects; advanced and trusted legal, regulatory, market and other institutions; stable political institutions; respect and trust internationally; low population density; geological stability; participation in the nuclear fuel cycle through a large resource endowment of uranium and a quite new, world-leading research reactor used for medical isotope production and international scientific research work.

Security and submarines

Reflecting on the conclusions reveals two 'loops'. One looks back on the 20th century, and the other spirals up to the present and beyond to the future. The first 'loop' is about reactor scale. The second 'loop' is about the naval origins of pressurised water reactor (PWR) technology, particularly its use for submarine propulsion.[45] These loops provide some perspective in looking forward and considering what sort of country Australia could be in the 21st century, and what sort of country it should be. As all people and nations must, Australia should first appreciate its legitimate national self-interest, the part that energy already plays and should play in the answer, the extent to which sound economics can be a guide, and whether or how our thinking may need to transcend the bounds of economic thinking alone.

Beyond the inward-looking perspective, we need to look outward and reflect on our endowment and our inheritance and ask: what roles should Australia rise to fulfil as a small, open, maritime trade-exposed global top-20 economy in the world's most dynamically growing region? What obligations do we have as inheritors of Westminster parliamentary democracy? What responsibilities do we have as a stabilising middle power in the Indo-Pacific neighbourhood? What expectations do our neighbours have of us as stewards of some of the world's largest deposits of uranium and of hydrocarbon resources, as well as vast tracts of geologically stable, arid lands topped up with super-abundant sun and wind and surrounded on all sides by oceans? What obligations and opportunities arise from Australia's very large export of higher education—not merely the revenues, but the large and growing cohort of students, scientists, engineers and professionals both in Australia and dispersed throughout the Indo-Pacific region?

Such questions have substantial economic aspects. They also transcend economics to the domains of national and energy security and environmental stewardship. These and other questions need to be weighed and considered alongside questions about how Australia is to meet its current and any future commitments under the Paris Agreement on climate change. Modularity, economies of replication, factory fabrication and series production as keys to technological learning represent an alternative path to 'gigantism' or the single-minded pursuit of economies of scale. These concepts are not unique to new engineering schools of thought for nuclear reactor design. This tack in engineering and technological development is part of the general phenomena of decentralisation that may be observed in everything from the desktop computer and mobile device revolutions to solar panels.

The modular revolution in reactor engineering design philosophy closes an interesting loop: SMRs such as the NuScale design are on the same scale as the first commercial nuclear power plant in the United States: the Shippingport 60 MWe demonstration reactor, which operated from 1957 to 1982 in Pennsylvania. At the same time, this 'return' to the small scale of early reactors embodies major advances for the future. SMR designs distil insights and lessons from decades of experience and technological developments, improving safety levels by orders of magnitude on the original small reactors. Having noted that potential commercial applications for nuclear energy extend beyond electricity generation to water desalination, hydrogen production, and industrial heat, we should not overlook marine applications. Indeed, PWR technology had its origins in naval (particularly submarine) propulsion. This is another important loop from the past to the present that also points to the future.

The change from diesel to nuclear propulsion of submarines that dates from the 1950s was revolutionary. The operational advantages and strategic consequences were more profound than Winston Churchill's far-sighted decision as First Lord of the Admiralty in 1911 to promote the view of the First Sea Lord, Admiral Sir John Fisher, that the Royal Navy switch from coal to oil.

Churchill, as first lord of the Admiralty, was the great champion of the Royal Navy's switch to oil, proclaiming, 'Mastery itself was the prize of the venture.' But in the years running up to the war, he encountered great opposition from both the Royal Navy and Parliament. Relying on oil, traditionalists argued, would leave the Royal Navy dependent not on its traditional, reliable supplies of Welsh coal but on oil from a highly unstable and unreliable nation: Persia (now Iran). In response, Churchill declared what would become the enduring principle of energy security: diversification. 'Safety and certainty in oil lie,' he said, 'in variety and variety alone.'[46]

A blue water navy with only diesel-battery electric submarines is at a profound strategic disadvantage to an adversary with nuclear propelled submarines. It is also highly constrained in its ability to operate with allies whose submarines are nuclear powered. The development in Australia of a fleet of commercial SMR PWRs alongside a naval fleet of submarine PWRs is probably not an essential pre-requisite but is likely to provide significant synergies and economies in operation and maintenance, in the workforce of engineers and technicians, and of the supporting facilities and infrastructure required to service and support each fleet.

Based on the assumption that the Royal Australian Navy (RAN) would need the capability to service its own vessels in Australia, the deployment of either a fleet of commercial nuclear reactors, or of nuclear propelled naval submarines, or both, would require the repeal of the various Commonwealth and state government laws that generally prevent deployment of nuclear reactors and related technology in Australia.

Repeal of the relevant portions of legislation is a small step in creating real options for deployment: it is necessary, but far from sufficient. Australia could choose to adopt a narrow perspective and approach energy policy in a silo, considering only economics, emissions and engineering reliability. Before doing so, the nation should at least consider whether a more holistic, strategic and comprehensive approach taking into account security, defence, national technological capabilities, and regional and global responsibilities would better serve the public interest.

✷ ✷ ✷ ✷

Units and terms

J   One joule, the metric **unit of energy** in the *System Internationale* (SI) is equal to the work done by a force of one newton (N) acting through a distance of one metre (m). In electrical terms, the joule (J) equals one watt-second (Ws)—i.e. the energy released in one second (s) by a current of one ampere (A) through a resistance of one ohm (Ω). One calorie is approximately 4.2 Joules.[47]

kJ   One kilojoule equals 1000 J.
A moderately active 80 kilogram 40 year old man will require food with about 13,500 kJ per day.[48]

MJ   One megajoule equals 1000 kJ. The above example is the same as 13.5 MJ.

GJ   One gigajoule equals 1000 MJ. Gas is priced in GJ on retail bills and in the wholesale market.[49]

W: One watt, the SI **unit of power**, equal to the release or conversion of one Joule per second (J/s)

kW   One kilowatt equals 1000 W

MW   One megawatt equals 1000 kW: the largest offshore wind turbine in the world, the Haliade-X from GE, which is 260m high with a 220m diameter rotor, is 12 MW. SMR nuclear plants are defined as having reactors between 50 and 300 MW electrical capacity.

GW   One gigawatt equals 1000 MW or 1 million kW: the scale of a large power station such as the brown coal-fired Loyang B plant in the Latrobe Valley, Victoria: has 2 units of 500 MW each. Large nuclear plants typically have reactors of about 1000 MW each.

kWh   One kilowatt-hour, the energy transferred by 1 kW of power operating for 1 hour, equal to 3600 J. The units in which retail electricity is priced (typically between 20 and 30 ¢/kWh)

MWh   One megawatt-hour equals 1000 kWh, the energy transferred by 1 MW for one hour, equal to 3.6 GJ.

GWh   One gigawatt-hour equals 1000 MWh. Annual output from a power plant is expressed in GWh.

TWh   One terawatt-hour equals 1000 GWh. Annual electricity generation and consumption by Australia may be expressed in TWh

Life: Power plants are assets with nine lives: financial asset life, design licence, plant licence, operator licence, site licence, technical asset life, refurbished or extended asset life, decommissioned site life, legacy issues and waste product lives. Beyondn the simple analysis in this chapter all types of lives require careful consideration in the development, financing, and deployment of nuclear plants.

Discounting, discount rate, cost of capital: the time value of money Discounted cash flow (DCF) analysis is used to transform a capital cost to its annualised equivalent.

The DCF method is at the core of LCoE calculations, as well as long-run marginal cost (LRMC) calculations.

Visualising twelve orders of magnitude

People: One person relative to **140 times** all the people currently alive on the earth

Money: one dollar to one trillion dollars: the coin in your pocket relative to the magnitude of the entire Australian economy (about $1.6 Trillion

Mass: One gram relative to one megatonne (million tonnes) or one kilogram relative to one billion tonnes. Australia exports more than one billion tonnes of iron ore each day, mainly to China.

Power: A 0.55 milliWatt CR2032 button cell weighs 3g, compared with a 550 MegaWatt multi-unit coal, or gas power station or a mid-size nuclear reactor. A new button cell has 235mAh or 653 mWh of energy, discharging a current of 0.19mA @2.9V, using $P = VI = 2.9 * 0.19 \times 10^{-3} = 0.551$ mW of power.

Time: one millisecond compared with 30 years or 1/100th of a second compared with 300 years.

10^{-3}	1 millisecond or 1/1000th of a second
10^{-2}	10 milliseconds (half an AC cycle @50Hz) 1/100th of a second
10^{-1}	0.1s = 100 milliseconds or 1/10th of a second (five AC cycles)
10^{0}	1s
10^{1}	10s
10^{2}	100 sec = order of 1 minute
10^{3}	1000 seconds = order of 15 minutes
10^{4}	10,000 seconds = 2.78 h ~ order of 3 hours
10^{5}	100,000 seconds = 27.8 hours ~ order of 1 day
10^{6}	1 million seconds ~ 10 days (order of week to a fortnight)
10^{7}	10 million seconds ~ 100 days (order of a quarter or a season)
10^{8}	100 million seconds ~ 1000 days (order of three years)
10^{9}	1 billion seconds ~ 30 years (third of a century)
10^{10}	10 billion seconds ~300 years (three centuries)

Energy:

0.2 kWh is the energy consumed by a 5W LED bulb in a desk lamp for 40 hours a week

2 kWh is the energy consumed in 1 hour by a 2kW fan heater or oil-filled radiator set to max

20 kWh is one day of electricity consumption for a typical home using 7300 kWh /y

200 kWh is less than half of the annual standby appliance consumption in average Australian homes.[50]

2 MWh is 2000 kWh, the energy per year for two AC units in a warm to hot climate

20 MWh is 20,000 kWh the energy consumed by a very large house in a year

200 MWh is about the hourly consumption of a 100,000 tonne potline in an aluminium smelter

2 GWh the energy generated in one hour by a large power station

at maximum output

20 GWh is close to one hour of average consumption in the National Electricity Market

200 GWh approximate scale of the annual energy consumption of the University of Queensland

2 TWh is about one percent of the energy consumed in the National Electricity Market each year

20 TWh is the energy generated by a large (2500 MW) plant operating continuously all year

200 TWh is approximately the electricity generated in the National Electricity Market each year

Endnotes

1 Values quoted in this chapter are in Australian dollars, unless otherwise specified. Paul Graham, Jenny Hayward, James Foster, Oliver Story and Lisa Havas, 2018, *GenCost 2018: updated projections of electricity generation technology costs*, CSIRO, Australia.

2 'The report then assigns a surprisingly high estimated cost to SMRs of $16,000 AUD/kW, as well as assuming almost no learning rate.' World Nuclear Association, *Submission 259*, p. 7, cited in *Not without your approval: a way forward for nuclear technology in Australia*: Report of the inquiry into the prerequisites for nuclear energy in Australia, House of Representatives Standing Committee on the Environment and Energy, December 2019, Canberra, p. 112. 'The GenCost 2018 capital estimates for nuclear power were a point of contention, with many inquiry participants expressing concerns and reservations about their validity...' para. 4.59 on p. 76 of the NSW Parliamentary inquiry report: Hon. Taylor Martin, MLC, Chair, New South Wales Parliament Legislative Council Standing Committee on State Development, Report no. 46, *Uranium Mining and Nuclear Facilities (Prohibitions) Repeal Bill 2019*, Sydney, March 2020, ISBN 9781920788599; The Parliament of Victoria, Legislative Council Environment and Planning Committee, *Inquiry into Nuclear Prohibition*, is continuing at the time of writing, with two hearings scheduled for August 2020. Transcripts page: www.parliament.vic.gov.au/epc-lc/article/4349.

3 The phase 'too cheap to meter' was first used in *Remarks Prepared by Lewis L. Strauss, Chairman, United States Atomic Energy Commission, For Delivery At The Founders' Day Dinner*, National Association of Science Writers, September 16, 1954, New York, p. 9. https://public-blog.nrc-gateway.gov/2016/06/03/too-cheap-to-meter-a-history-of-the-phrase/. *The Economist*, 'Nuclear Power: Atomic Renaissance', 6 September 2007 is probably the first use of the quip 'too expensive to matter', which has gained more common use recently.

4 Paul Graham and Jenny Hayward, 2017, *Electricity Generation Technology Cost Projections: 2017–2050*, CSIRO, Australia.

5   Cole Latimer, 'Australia has 'missed the boat' on nuclear power', *Sydney Morning Herald*, 11 January 2018.

6   Paul Graham, Jenny Hayward, James Foster and Lisa Havas, *GenCost 2019–20: preliminary results for stakeholder review*, CSIRO, Australia.

7   Tom Mundy, *NuScale SMR Progress to Deployment*, Submission to the House of Representatives Standing Committee on the Environment and Energy, August 2019. Tom Mundy of NuScale provided further information on the basis for these costs, which are overnight capital costs, first estimated in 2014 and updated in 2017, in an email dated 27 August 2020 in response to questions on notice from the Victorian Inquiry into Nuclear Prohibition, https://www.parliament.vic.gov.au/epc-lc/article/4349

8   DER is defined as '[i]ncluding rooftop PV, batteries, and other resources at the customer level.'
VRE is defined as '[i]ncluding solar, wind, and other variable renewable energy resources at the utility level. New dispatchable resources refers to storage using pumped hydro and batteries, as well as gas-fired peaking plant. Audrey Zibelman, *2020 Integrated System Plan for the National Electricity Market*, AEMO, 30 July 2020, p. 9: https://aemo.com.au/energy-systems/major-publications/integrated-system-plan-isp/2020-integrated-system-plan-isp

9   The context in Australia (and other mature, industrialised economies, particularly in the OECD) contrasts with the situation of the rapidly emerging, developing and advancing economies, especially across Asia, where the power generation fleets are far younger, and where the unavoidable decisions centre on investments to meet growth.

10 The PESTLE mnemonic—political, economic, social, technological, legal and environmental—is classically used to group the diverse range of factors. 'Commercial' is used here for economic, financial and market-related factors.

11 Audrey Zibelman, 2020 *Integrated System Plan for the National Electricity Market*, AEMO, July 2020, p.44 .

12 Mr Koji Inoue, Director-General, Natural Resources and Energy Policy, Ministry of Economy, Trade and Industry, Japan, opening remarks commenting on Japan's New Strategic Energy Plan, 35th Australia-Japan High Level Group Consultation on Energy and Mineral Resources, Brisbane, 23 June 2014.

13 The National Electricity Market 'regions' are approximately coincident with state borders, but not completely. The Australian Capital Territory is part of the NSW market region, and there are some quirks in the configuration of the network in a few places near state borders.

14 The State of the Energy Markets from the AER provide detailed data, information and explanation.
https://www.aer.gov.au/publications/state-of-the-energy-market-reports

15 In Australia in 2019, just over 265 000 GigaWatt-hours (GWh) of electricity were generated, of which a little over 188 000 GWh were generated on the National Electricity Market in Eastern Australia.

16 AEMO, 2020, *Renewable Integration Study: Stage 1 Report*, p. 6.

17 Australian Energy Regulator, 2018, State of the Energy Markets, Table 2.4, p. 98.

18 Strictly speaking, the data here is half-hourly generation, which is equal to consumption plus energy losses throughout the transmission and distribution networks. In other words, it is electrical demand as measured at the busbar of the power plants, rather than electrical demand summed up across all customers' meters.

19 https://aemo.com.au/energy-systems/electricity/national-electricity-market-nem/data-nem/aggregated-data.

20 Australian Energy Regulator, 2018, lists the portfolios of generators by market region, and the holder of the trading rights, showing also the plant capacities and the owners, whether private or government.

21 https://www.aemc.gov.au/regulation/energy-rules/national-electricity-rules.

22 AEMO, 2020, describes many of these challenges, along with appendices as separate volumes on frequency control, and managing variability and uncertainty.

23 Gabriel Rioseco, 2019, PhD Confirmation Report, based on 14 from more than 50 papers from literature review of VRE Integration costs, supervised by the author.

24 This effect can already be observed in South Australia, and is one of the reasons for the higher prices in that region. Gas-fired plants must operate to maintain the system in a secure state, even when wind and solar generation within the region exceeds South Australian demand. As a consequence, the incremental gas-fired generation is 'forced in', effectively moving from the top of the short-run marginal cost curve (and further up the bid stack) to a 'must-run' status, and the energy flows eastwards towards Victoria. In market terms, subject to transmission constraints between Victoria and NSW, the effect is probably to displace black coal-fired generation in NSW, which has higher variable costs than the brown coal-fired generation in Victoria. See the 'Case study | Minimum operational demand in South Australia' in AEMO, 2020, *Renewable Integration Study: Stage 1—Appendix A: High Penetrations of Distribution Solar PV*, pp. 37–38.

25 AACE International Recommended Practice No. 18R-97 (2016) *Cost Estimate Classification System—As Applied In Engineering, Procurement, and Construction for the Process Industries*.

26 The controversial estimates in the CSIRO report provide neither ranges nor specify the estimate class or the use.

27 See, for example: ACCC, *Restoring Electricity Affordability and Australia's competitive advantage: Retail Electricity Pricing Inquiry—Final Report*, Commonwealth of Australia, June 2018, Canberra. Figures A and B on pp. v and vi.

28 This was recognised at least as early as 1995, but the United States National Renewable Energy Laboratory. See the selection criteria guide in chapter 3 by Walter Short, Daniel J. Packey, and Thomas Holt, 1995, *A Manual for the Economic Evaluation of Energy Efficiency and Renewable Energy Technologies*, NREL.

29 This change has no effect on the result, because the calculations are driven from the electrical capacity, not the thermal capacity.

30 Fuel plus all operating and maintenance costs are estimated as about $A24/ MWh. From personal communication with Tony Irwin, former ANSTO engineer and Technical Director of SMR Nuclear Technology. Tony is probably the most experienced nuclear engineer in Australia, having commissioned eight of the nuclear reactors in Britain and, the new OPAL reactor at Lucas Heights, which was delivered on-time and on-budget and has been operating at high utilisation rates since.

31 Data and statistics on the electricity and non-electricity parts of Australia's energy sector, and the domestic, export and imports of energy are available at the Department of Industry, Science, Energy and Resources: https://www.energy. gov.au/data-charts; https://www.energy.gov.au/publications/australian-energy-statistics-table-o-electricity-generation-fuel-type-2018–19-and-2019.

32 Ziegler, et al, *A Compendium for a sensible energy policy*, Bundesinitiative Vernunftkraft, November, Berlin, 2017.

33 A framework for thinking about the deployment of technologies into 'ecosystems' is provided by. Adner and Kapoor (2016) 'Right Tech, Wrong Time,' *Harvard Business Review*, November. https://hbr.org/2016/11/right-tech-wrong-time. This chapter discusses the specifics of deployment within a power system with a competitive spot market.

34 https://www.theguardian.com/environment/2020/mar/12/ wind-and-solar-plants-will-soon-be-cheaper-than-coal-in-all-big-markets-around-world-analysis-finds.

35 The load-duration curve is simply the demand for electricity in hourly (or half-hourly or five-minute) time intervals during a year, sorted from highest to lowest. In classical system planning (the way the Eastern Australian grid was planned and developed for decades prior to privatisation), there is a large amount of plant with high fixed costs and low variable costs to meet the invariant portion of demand. This graphical method was used before computing power was sufficient for system simulation. It loses the chronological effects, but in a traditional system should give a good approximation to a more detailed computer simulation. The method remains useful for explaining the economics of generation systems and why almost every system in the world has a fuel mix.

36 ACCC, 2018.

37 https://aemo.com.au/energy-systems/electricity/national-electricity-market-nem/ data-nem/aggregated-data.

38 This finding comes from primary research interviews conducted in June 2017 for a confidential client by the author and Professor Chris Greig, of almost all of the major Australian, American, Asian and European banks active in lending to energy, resources and infrastructure projects in Australia.

39 Paul L Joskow and Richard Schmalensee, 1983, *Markets for Power: An Analysis of Electric Utility Deregulation*, MIT, p.133. The authors discussed their old book at length in an interview in June 2019. RS: I went back and looked at the book a few months ago. It was all about what kind of long-term contracts. You need long-term contracts to finance new capacity: what kind, who would do them. There's

a sentence, which I could probably find again, where it says, we doubt the ability of market prices to provide adequate guidance for investment decisions. Nobody else doubted that, really, for a while. But I think we're back to the future, in a way. If you look at that book, it says we need intelligent decisions about the capacity mix and long-term contracts to enable financing. Here we are. PJ: I think maybe the worst thing is to back into this with your eyes half-closed. If we're going to a system where we're basically doing integrated resource planning again, we're going to be deciding on wind and solar and batteries and quick start and fast response generators and we're going to be contracting those up to get them and keep them in the market. Maybe we should just recognize that and start putting into place mechanisms to make a system like that work efficiently, rather than waiting until we've driven all of the incumbent capacity out of the market, then trying to figure out how we're going to make it work. Francis O'Sullivan, *Electricity Markets*, MIT Energy Initiative Podcast #14, http://energy.mit.edu/podcast/electricity-markets/.

40   *Renewable Energy (Electricity) Act*, 2001.
41   https://www.world-nuclear.org/information-library/facts-and-figures.aspx.
42   International Energy Agency, 2003, *World Energy Outlook*, Paris, p. 429.
43   Hyman G. Rickover, *Memo*, Naval Reactors Branch, Division of Reactor Development, United States Atomic Energy Commission, 5 June 1953. Admiral Rickover built the first full-scale commercial demonstration reactor in the United States at Shippingport, commenced 6 September 1954, commissioned 26 May 1958 (45 months) and decommissioned December 1989.
44   Examples include the Monroe Doctrine of 1823, the Roosevelt Doctrine as initially set out in the *Atlantic Charter*, the Truman Doctrine on containment of Soviet spread, the Eisenhower Doctrine on the prevention of armed aggression, the Kennedy Doctrine on the defence of liberty, the Reagan Doctrine that the Cold War was winnable, the Bush Doctrine on advancing liberty and hope.
45   Norman C Polmar and Norman Friedman, 'Submarine: Nuclear Propulsion,' *Encyclopaedia Britannica*, 10 June 2019, www.britannica.com/technology/submarine-naval-vessel/Nuclear-propulsion.
46   Daniel Yergin, *The Prize: The Epic Quest for Oil, Money and Power*, 1990.
47   https://www.britannica.com/science/joule.
48   https://www.betterhealth.vic.gov.au/tools/energy-needs-kilojoules-calculator.
49   Australian wholesale gas prices can be viewed at https://aemo.com.au.
50   https://www.originenergy.com.au/blog/understanding-standby-power/.

9   Nuclear power and climate change

Ben Heard

Rapid disruption of the Earth's climate is already well underway. In 2019, the global average temperature was estimated to be 0.95° celsius above the 1950–1981 average.[1] Best-estimate modelling suggests that temperatures could rise to between 2.6 and 4.8° celsius above the 1986–2005 average by the end of this century.[2] Emissions of greenhouse gas are driving this rapid warming. Present energy consumption, based almost entirely on the combustion of fossil-carbon fuels, will result in the emission of the long-lived greenhouse gas, carbon dioxide. The combustion of oil, coal, and gas provides us with electric power, heat and transportation, and emits 33 billion metric tonnes per year of carbon dioxide. Beyond the (substantial) inputs of fossil energy, our food production systems are also a major source of greenhouse gas, through the production of fertiliser, changes in land use, and emissions from livestock. Even when energy itself is clean, critical industrial processes for materials like steel and cement are major direct sources of emissions.

That we can alter our climate is inarguable. We must now act with a controlled sense of urgency to resume control over this process. For while the adaptive capabilities of humans are impressive, such rapid increases in global average temperature (and changes in ocean chemistry) might vastly exceed these capabilities. We are presently on a trajectory where we will experience a climate that has not existed at any time in the journey of our species. Our most basic sense of conservatism and risk management demands an attentive, concerted response.

It is perhaps understandable why some grasp at reasons to look away. This challenge runs through every aspect of our civilisation, an unintended consequence of our amazing ability to shape the world around us in the service of human prosperity. Furthermore, it cannot be resolved through incremental 'improvements' in the form of lowering 'emissions intensity' (less emissions per unit of economic output) that leave annual emissions growing, nor even slow and steady reduction of annual emissions to some more psychologically acceptable level. Unlike air-pollution, where an environmental and social benefit is realised quickly and permanently with the reduction of the polluting source, greenhouse gas emissions are long-lived, accumulating in the atmosphere to drive ever-greater climate-related risks. The daunting target that must be achieved, and then maintained in perpetuity, is near-zero emissions, from all sources, in all places, for ever.

Consequently, humanity faces a seemingly paralysing paradox. How can we continue to enjoy the benefits of a modern, energised civilisation, extending it to eliminate poverty and offer pathways to realise human potential, without fatally undermining that most vital of all our natural support systems—a stable, dependable and hospitable climate? If this notion of a 'paralysing paradox' seems exaggerated, consider our fortunes over the first three decades of concerted international effort in climate policy. In 1990, the Intergovernmental Panel on Climate Change produced its first assessment report. Across three decades of increasing scientific understanding of, and policy focus on, anthropogenic impacts on the climate, the proportion of global primary energy supplied by fossil fuels has *increased* from 84 per cent to 86 per cent.[3] Energy use has appreciably altered in one major way: it has grown, along with increasing human population and rising standards of living.

Between 1990 and 2019, the non-hydro renewable contribution to electricity generation had grown 23-fold (to 2,805 TWh). Yet, total global electricity consumption more than doubled during that same period (11,914 to 27,005 TWh). All other non-electrical energy (principally heat and transportation) remained dominated by fossil fuels. There has been no transition away from fossil fuels of sufficient magnitude to offset growth, with annual carbon dioxide emissions from energy increasing 60 per cent from 1990–2019, rising 1.1 per cent annually for the decade to 2019.

According to the framing of political scientist Jon Symons, the developing world's principle response to climate change for the last 30 years has been 'a form of adaptation: rapid economic growth' which has 'dramatically improved the resilience of entire countries', reducing the risk of exposure to climate harms—harms that were already real for those in deep poverty.[4] To deny humans this path would be a monstrous injustice. As Indian economist Samir Saran frames it: 'our poverty must not be your mitigation strategy'.[5] Evidence from the previous three decades shows that humanity is unwilling to forgo its access to energy today in order to mitigate the shared, less-tangible future risks presented by climate change. This is entirely rational for the global poor, who derive greater safety and well-being, including from exposure to climate related hazards, through more stable dwellings, healthier water supply, greater food security, and improving health care. Most simply put, given the choice between dirty energy and inadequate energy, dirty energy wins every time.

Addressing climate change decisively is a challenge of the most extraordinary scale, and three decades of experience asserts that a low-energy world is a strategy devoid of foundation. Furthermore, the encouraging adage that 'small actions can make a big difference' is sadly mistaken when dealing with this particular dilemma. Climate change mitigation requires large actions to make large differences. The world needs a plausible pathway for the permanent, deep de-carbonisation of global economic activity that does not hinge on denying moderately prosperous lives to other people. To the greatest extent possible, this ought to be treated as an engineering and innovation challenge. Let our political disagreements continue against consensus for action: that we can disagree just as effectively in a de-carbonised world.

This chapter examines the role of nuclear power in that deep de-carbonisation challenge. I will explore how nuclear power offers a pathway to solutions of relevant scale for the major sources of greenhouse gas emissions. As I will show, the solutions return to the availability of plentiful, dense, reliable and affordable energy that can be delivered without greenhouse gas emissions.

Electrical power generation

A logical starting point is electrical power generation—the main use of modern nuclear technology. In generating nuclear energy, humanity developed the first, and still only fuel-based energy source that does not rely on the process of combustion (rapid oxidation) of carbon-based fuels. Nuclear technologies deploy the wholly different physical process of fission. The difference is not merely qualitative but quantitative as well. Human civilisation advanced with the exploitation of fuels of higher energy density. Where dry firewood holds around 16 MJ kg^{-1}, good quality coal has nearly double the density (30 MJ kg^{-1}) and crude oil approximately triple the density (45–46 MJ kg^{-1}). Natural uranium, deployed in a typical light-water reactor, offers around 500,000 MJ kg^{-1}—an energy density *five orders of magnitude greater than crude oil.*[6] This opens avenues for unexpected and beneficial progress in climate mitigation through the applications of this extraordinarily portable, dense fuel.

With these properties, nuclear fission is the only form of greenhouse-gas free energy production that has been proven and demonstrated beyond doubt as reliable, fully transferable and completely scalable to the electrical power demands of developed nations. It is the second largest source of clean power production in the world, and the largest source of clean power production in both the United States and the European Union. The global nuclear power sector has been assessed as responsible for avoiding 64 GtCO^2-e, as well as preventing in the order of 2 million deaths through avoided air pollution.[7]

In nations where nuclear has been deployed in decades past, it has proven potent in displacing fossil fuels from electricity supply, with electricity costs that are generally low and stable to the extent that in Sweden a tax on nuclear electricity made up one-third of the operating cost.[8] The role of nuclear power in electricity production is discussed in greater detail elsewhere in this book. I would simply note here that 55 reactors are under construction world-wide, several nations are developing nuclear power for the first time, and innovation is proceeding rapidly to bring a greater variety of nuclear power options to serve broader markets. From a technology and engineering perspective, the case for nuclear power in fulfilling a need of this magnitude is unassailable. The principal global challenge is driving down costs and raising enthusiasm for nuclear solutions. In an affluent nation like Australia, government policies that arbitrarily prohibit the possible introduction of nuclear technology represents poor global citizenship.

Electrified transportation

While there is no obvious line of sight on privately owned fission-mobiles, clean reliable electricity at scale cannot be ignored as part of the transition to electric vehicles. Australia's 'Integrated System Plan 2020' suggests the demand for vehicle charging to be somewhere between 17–40 TWh per year of electricity by 2050.[9] This much power alone would require the output of several thousand megawatts of reliable production, such as that provided by nuclear technologies. A wholly de-carbonised grid, blending both renewable and nuclear technologies, can facilitate uptake and maximise the benefits of electric vehicles (EV). Owners would not be constrained by time or charging conditions based on weather. Clean power would simply be available around the clock at any time of year. Smart integration of EV charging might provide additional network and consumer benefits, but the beneficial uptake of EVs themselves by millions of private consumers ought not, and need not, depend on this smart integration for the climate mitigation outcome. With a grid based on the right blend of technologies, an EV can be charged when convenient.

Nuclear technologies for process heat

Process heating supplies thermal energy to transform materials into a wide variety of industrial and consumer products, including ubiquitous materials (such as concrete and steel), chemicals (such as hydrogen and ammonia), and processed food.[10] The sector is heterogeneous although there is scant data offering accurate breakdowns of requirements by both quantity and temperature.[11] For example, food processing demands temperatures from 65–250° celsius. Some common chemical processes such as the production of hydrogen or ammonia demand 500–1,000° celsius. The smelting of metals and the processing of metal ores applying calcination and hardening can demand 800–1500° celsius. Industrial-scale heating requirements worldwide are almost entirely met through combustion of fossil gas and coal.

There are evident limitations to the scalability of renewable resources for lower temperature process heat, and no foreseeable, cost-effective options for providing higher temperatures.[12] A limited amount of need might be met with electrification. But this need raises demand for clean electrical power and competes with existing de-carbonising requirements. No matter how much the cost of a solar panel decreases, it is a weak pathway to the provision of constant, high-grade heat.

Fortunately, a nuclear reactor provides continuous and reliable heat at industrial scale. Commercially mature light-water reactors offer relatively low outlet temperatures (approximately 350° celsius), suitable for uses up to lower temperature industrial requirements. Much development in the advanced nuclear sector is focussed on reactors that are smaller, with outlet temperatures ranging from 500–1,000° celsius, suitable for a range of industrial applications.[13] For the goal of deep de-carbonisation, nuclear technologies are arguably indispensable for meeting the industrial heat challenge.

Nuclear propulsion

While privately owned fission-mobiles might be some way in the distance, nuclear technologies already rule the ocean in terms of de-carbonisation. Nuclear power is a proven solution for ocean-going propulsion, being employed in submarines, some naval vessels and ice-breaking vessels. The potential benefits from expanding these uses to include cargo ships is too great to be ignored. Shipping alone accounts for 2.5 per cent of global emissions, as much as Japan's domestic output, and is forecast to climb 250 per cent by 2050.[14] It is a major, not minor, portion of the mitigation task demanding a solution. The big difference is the immediate availability of a technology solution, entirely proven in small nuclear reactors that presently propel 140 vessels.

A move to broad civilian application naturally presents challenges, such as the need for re-designed vessels, specific crew training requirements and the spectre of social acceptance. It also brings potential benefits, including total fuel autonomy, saving time, money and reducing volatility in operating profits from varying fuel prices. Along with hydrogen, ammonia and advanced biofuels, it is one option identified by the International Transport Forum with the potential for 100 per cent reduction in CO^2 emissions.[15] The pathways to deliver that quantity of hydrogen and ammonia fuel that I will consider below are almost certain to require fission as part of their original energy source.

Novel applications—hydrogen, ammonia and synthetic fuels

One important application of advanced nuclear reactors is the greenhouse gas-free creation of hydrogen, which has myriad applications in a deep de-carbonisation framework.[16] With clean, reliable, low-cost heat and electricity, hydrogen can be produced from the ambient environment using high-temperature steam electrolysis.[17] This cleanly produced hydrogen could substitute for methane in ammonia production, offering huge de-carbonisation potential in the essential provision of secure food supplies for 10 billion people. Cleanly produced hydrogen can also be applied as the reductant in direct-injection iron production, resulting in higher-quality iron with much reduced greenhouse gas emissions when compared with blast furnace production.[18] Hydrogen can be combined with carbon-dioxide to create synthetic crude oil (C_nH_{2n+2}), methanol (CH_3OH), or dimethyl ether (C_2H_6O).[19] These chemicals are energy-dense, stable, easily stored and trans-ported and can be refined into the full range of hydrocarbon fuels. Where the carbon dioxide is re-used or sourced from the ambient environment rather than fossil sources, such fuels could be emissions neutral. While electrifi-cation holds huge potential for passenger transport and a limited range of freight requirements, the complementary development of these fuels could virtually eliminate net greenhouse gas emissions from transport and other processes that demand directly combustible fuel.

Novel applications—intensive agriculture
The conversion of land for agriculture is a significant source of greenhouse gas emissions. Part of the answer lies in intensification, using clean energy as a substitute for the products and services otherwise provided by nature in a more dilute form. Australia already hosts one of the world's best examples, Sundrop Farms near Port Augusta in South Australia. In this case, solar thermal energy is deployed for desalination, heating, ventilation, and nutrient production under greenhouse conditions. This farm delivers astonishing yields of nearly 500 tonnes per hectare, well more than the best perform-ing advanced farms in Israel, and orders of magnitude greater yield than low technology, field-based cropping. The land sparing, conservation and greenhouse benefits of energy intensive advanced agriculture are potentially huge. Nuclear technologies enhance that potential with an energy source that can be brought to any location, rather than relying on locations of excellent ambient resource.

Further into the future we have the potential for production of some (vir-tually) 'landless' staple crops, achieved by directly feeding micro-organisms with hydrogen. This is a more technologically radical view of progress, one that views farmland as akin to 'large and unwieldy surfaces for the inefficient collection of sunlight'.[20] The land-sparing and, thus, climate mitigation potential is even more appealing. Substituting just 2 per cent of the dry matter for animal feed with these microbial proteins might spare 100 million acres from agricultural conversion.[21] The catch continues to be finding a replacement clean energy. The 100-million acre outcome would require 300 GW worth of nuclear power driven electrolysis for hydrogen production. Food researcher Linus Bloomqvist insightfully places this level of production into the future but wisely notes that the real constraints to sustainability are 'not any inherent ecological carrying capacity, but rather technology, and the institutions that create and sustain it'.[22]

Novel applications—direct air carbon dioxide capture
Despite the climate commentariat's encouragement to keep striving for emissions reduction targets, just as many serious observers of emission trends and climate sensitivity contend that we are likely or certain to overshoot the global emissions budgets we might have estimated for either 1.5 or 2 degrees celsius of warming. This leaves the task of achieiving 'negative emissions'—withdrawal of carbon dioxide from the air to reduce concentrations and limit climate change.

'Negative emissions' technology occupies an interesting space in climate mitigation. It is indisputably a form of deliberate geoengineering, applying technology to knowingly and directly alter atmospheric chemistry to create preferred conditions. Unlike, for example, solar geo-engineering through the stratospheric distribution of short-lived aerosols, it attracts a fraction of the controversy. Perhaps 'negative emissions' exists in a psychological framing of 'correction', or righting a wrong and restoring a pre-existing equilibrium, whereas 'geo-engineering' exists in a framework of 'tampering'.

Regardless of framing, negative emissions through technological means is going to be an energy hungry process. According to a 2019 paper published in *Nature Communications*, scaling up Direct Air Capture And Storage (DACCS) to 30 Gt per year in 2100, even complementing other negative emission initiatives, requires additional primary energy of 225–250 Ej per year. This figure is greater than half of global total final consumption in 2016.[23] Implementing such energy-hungry mitigation, which might indeed prove necessary, provides a straightforward need for the scalable, reliable, energy dense zero-carbon heat and power that is offered by nuclear fission.

Summary

A holistic and unflinching view of climate mitigation suggests that the role of nuclear power in managing climate change will be broad, deep and multi-faceted. To date, we have made hard work on the least of these challenges, that of decarbonised power grids. This belies the suite of more challenging innovations, all requiring additional input of heat and power, that will be required to both eliminate greenhouse emissions from our economic activity, and achieve atmospheric concentrations of greenhouse gas commensurate with a stable and reliable climate.

Can the world find a way to universal prosperity and human dignity within a civilisation that has transitioned its relationship with the natural world from one of exploitation to restoration? Possibly. But this can only happen by merging ingenuity with abundant and affordable, low-carbon energy. Nuclear technologies are not the 'silver bullet'. They are merely indispensable. Rising to these challenges without the uniquely beneficial characteristics of nuclear technology is, at best, misinformed. At worst, it is hubristic folly. Efforts in deployment, innovation, research and development must be accelerated to achieve the breakthroughs in cost that will allow the deployment of nuclear technologies beyond merely replacing coal in existing grids. Nuclear power is the foundational energy source upon which deep de-carbonisation can be achieved. There is urgent need for a process of political and cultural normalisation. In the Australian context, the elimination of arbitrary legislative prohibitions is the most logical starting point.

★ ★ ★ ★

Endnotes

1   NASA, *Global Temperature*, NASA, https://climate.nasa.gov/vital-signs/
global-temperature/.

2   IPCC, 'Summary for Policymakers' in *Climate Change 2013: The Physical
Science Basis. Contribution of Working Group I to the Fifth Assessment Report of
the Intergovernmental Panel on Climate Change*, Cambridge University Press,,
Cambridge, 2013.

3   BP, *BP Statistical Review of World Energy*, BP, London, 2019.

4   Jon Symons, *Ecomodernism: Technology, politics and the climate crisis*, Polity Press,
Cambridge, 2019.

5   Symons, *Ecomodernism*.

6   World Nuclear Association, *Energy for the World - Why Uranium?*, World Nuclear
Association, http://www.world-nuclear.org/information-library/nuclear-fuel-
cycle/introduction/energy-for-the-world-why-uranium.aspx.

7   Pushker Kharecha and James Hansen. 'Prevented mortality and greenhouse gas
emissions from historical and projected nuclear power.' Environ Sci Technol, vol.
47, no. 9, 2013.

8   Friedrich Wagner and Elizabeth Rachlew. 'Study on a hypothetical replacement
of nuclear electricity by wind power in Sweden', *European Physical Journal Plus*,
vol. 131, no. 5, 2016. ; Staffan Qvist and Barry Brook, 'Potential for worldwide
displacement of fossil-fuel electricity by nuclear energy in three decades based
on extrapolation of regional deployment data', *PLoS One*, vol. 10, no. 5, 2015. ;
Steve Kidd. 'Nuclear power plants--how have they become like ATMs?', *Nuclear
Engineering International*, vol. 56, no. 679, 2011; Nicolas Boccard, 'The Cost of
Nuclear Electricity: France after Fukushima', Centre for Economic Policy Research,
Research School of Economics, Australian National University, 2013.

9   Australian Energy Market Operator, *2020 Integrated System Plan*, AEMO,
Australia, 2020.

10 United States Department of Energy, *TA 61: Industrial Process Heating Systems*,
Washington DC, 2015.

11 Tobias Naegler et al, 'Quantification of the European industrial heat demand by
branch and temperature level', *International Journal of Energy Research*, vol. 39, no.
15, 2015.

12 United States Environmental Protection Agency, *Renewable Industrial Process
Heat*, https://www.epa.gov/rhc/renewable-industrial-process-heat.; Keith
Lovegrove et al., *Renewable energy options for Australian industrial gas users*, IT
Power Australia, Canberra, 2015.; Naegler et al, 'Quantification of the European
industrial heat demand by branch and temperature level.'; C Lauterbach et al,
'The potential of solar heat for industrial processes in Germany.' Renewable and
Sustainable Energy Reviews 16, no. 7, 2012. ; Emanuele Taibi, Dolf Gielen, and
Morgan Bazilian. 'The potential for renewable energy in industrial applications',
Renewable and Sustainable Energy Reviews, vol. 16, no. 1, 2012.

13   Idaho National Laboratory, 'The high-temperature gas-cooled reactor next generation nuclear energy,' (Idaho Falls, Idaho: Idaho Naitional Laboratory, Undated) ; Zuoyi Zhang et al, 'The Shandong Shidao Bay 200 MWe High-Temperature Gas-Cooled Reactor Pebble-Bed Module (HTR-PM) Demonstration Power Plant: An Engineering and Technological Innovation', *Engineering*, no. 2, 2016.

14   Leigh Phillips, *If Shipping Were a Country Part 1*, The 4th Generation: News and Persepctives on the Future of Energy, https://4thgeneration.energy/if-shipping-were-a-country-part-1/.

15   International Transport Forum, *Deacarbonising Maritime Transport: Pathways to zero-carbon shipping by 2035*, OECD/ITF, 2018.

16   Marek Jaszczur et al, 'Hydrogen production using high temperature nuclear reactors: Efficiency analysis of a combined cycle', *International Journal of Hydrogen Energy*, vol. 41, no. 19, 2016; Adedoyin Odukoya et al, 'Progress of the IAHE Nuclear Hydrogen Division on international hydrogen production programs', *International Journal of Hydrogen Energy*, vol. 41, no. 19, 2016. ; Jean Leybros et al. 'Plant sizing and evaluation of hydrogen production costs from advanced processes coupled to a nuclear heat source: Part II: Hybrid-sulphur cycle', *International Journal of Hydrogen Energy*, vol. 35, no. 3, 2010.

17   Magali Reytier et al, 'Stack performances in high temperature steam electrolysis and co-electrolysis.' International Journal of Hydrogen Energy 40, no. 35, 2015 ; Heather D. Willauer et al, 'Feasibility of CO2 Extraction from Seawater and Simultaneous Hydrogen Gas Generation Using a Novel and Robust Electrolytic Cation Exchange Module Based on Continuous Electrodeionization Technology.' Industrial & Engineering Chemistry Research 53, no. 31, 2014; Matthew D Eisaman et al, 'CO2 extraction from seawater using bipolar membrane electrodialysis.' Energy & Environmental Science 5, no. 6, 2012.

18   Lionel Germeshuizen, 'A techno-economic evaluation of the use of hydrogen in a steel production process, utilizing nuclear process heat.' International Journal of Hydrogen Energy 38, 2013.

19   Emily O'Dowd, *Norway sees the biggest investment for Blue Crude yet*, https://www.biobasedworldnews.com/norway-sees-the-biggest-investment-for-blue-crude-yet.; George A Olah, Alain Goeppert, and GK Surya Prakash, *Beyond Oil and Gas: The Methanol Economy*, Wiley-VCH, Weinheim.

20   Calder cited in Linus Blomqvist, *What Microorganisms Can Teach Us about Decoupling and Limits to Growth*, The Breakthrough Institute, https://thebreakthrough.org/issues/conservation/microorganisms.

21   Ilje Pikaar et al, 'Decoupling Livestock from Land Use through Industrial Feed Production Pathways', *Environmental Science and Technology*, vol. 52, no. 13, 2018.

22   Linus Blomqvist, 'What Microorganisms Can Teach Us about Decoupling and Limits to Growth'.

23   Giulia Realmonte et al, 'An inter-model assessment of the role of direct air capture in deep mitigation pathways.' Nat Commun, vol. 10, no. 1, 2019.

Part Three

10 Australia's strategic outlook and submarine capability

Malcolm Davis

*A*ustralia's strategic outlook is more dangerous and contested as our strategic gaze shifts towards the 2030s and beyond. The 2020 Defence Strategic Update highlights growing strategic competition between the US and China as the principal driver of strategic dynamics in our region, with this competition playing out across the Indo-Pacific region. It notes that: 'since 2016, major powers have become more assertive in advancing their strategic preferences and seeking to exert influence, including China's active pursuit of greater influence in the Indo-Pacific.'[1]

The Defence Strategic Update also notes that states are accelerating their military modernisation with 'the introduction of advanced strike, maritime surveillance and anti-access and area denial capabilities, [with] ... new weapons being introduced into the region [that] have increased range, speed, precision and lethality.'[2] It highlights a growing challenge from emerging and disruptive technologies that could see 'sophisticated sensors, autonomous systems and long-range and high speed weapons' introduced into our region.[3]

In launching the 2020 Defence Strategic Update and the companion Force Structure Plan on 1 July 2020, the Prime Minister, Scott Morrison, drew a parallel between the deteriorating strategic future facing Australia in the next decade, and the tumultuous years of the 1930s.

> We have been a favoured isle, with many natural advantages for many
> decades, but we have not seen the conflation of global, economic and

strategic uncertainty now being experienced here in Australia in our region since the existential threat we faced when the global and regional order collapsed in the 1930s and 1940s.[4]

The Minister for Defence, Senator Linda Reynolds, explained: 'major power competition, militarisation, disruptive technological change and many other new threats are all making our region less safe'. She made clear that notions of a 'ten-year strategic warning period' were no longer valid as a basis for Australian defence planning.[5]

With this caveat in mind, our choices regarding the Royal Australian Navy's future submarine project need to be benchmarked against this more adverse future. More specifically, we need be prepared for the 'less remote' possibility of major power interstate war between China and the United States that would likely involve Australian commitment of military forces under the ANZUS Treaty. With this prospect in mind, preparing for major power interstate war means sustaining—and expanding—Australia's military capabilities. The 2020 Defence Strategic Update and Force Structure Plan focus on just this issue but, on the issue of submarines, it remains committed to the future acquisition of 12 *Attack* class conventional submarines from 2034. Nuclear powered and propelled submarines are not considered in this most recent policy announcement by the Government.

The 2016 Integrated Investment Program (IIP) noted that a submarine review will be conducted in the late 2020s to consider whether the configuration of the submarines remains suitable or whether 'consideration of other specifications should commence'.[6] The 2020 Force Structure Plan does not mention this review although the Government remains committed to the *Attack* class future submarine as a 'regionally superior' submarine capability.[7] This term is never explicitly defined by the Government or by the Royal Australian Navy (RAN).[8] The 2016 Defence White paper also suggests that '...by 2035, around half of the world's submarines will be operating in the Indo-Pacific region where Australia's interests are most engaged'.[9] Yet, none of the ASEAN states are intending to acquire such a large number of submarines as to be able to operate 'half the world's submarines' under their current defence planning.[10] It seems more likely that the submarine forces of China need to be considered if this statement is to be meaningful.

I will examine Australia's strategic outlook, with a focus on China's evolving naval capability, the risk of that capability being deployed closer to Australia's maritime approaches, and the potential transformation of the future undersea battlespace. It makes the case that given the emerging strategic outlook, Australia should transition from conventional *Attack* class submarines to much more capable nuclear submarines in the future, and invest more readily in development of advanced unmanned underwater vehicles, along with sophisticated undersea surveillance capabilities.

Australia's dangerous strategic outlook

Major power competition and high-intensity interstate war between the United States and its authoritarian peer adversaries—China and Russia—is once again first and foremost on the minds of many strategic analysts. [11] Robert Kaplan argued in 2005 that 'the American military contest with China ... will define the twenty first century' and that this Cold War would predominantly be fought within a maritime context, with naval power a key tool of both sides to determine the outcome. [12] He then noted in 2019 that the future he predicted in 2005 has arrived and makes clear that China is the *pacing threat* the US Military now measures itself against. [13] He argued that

> the Chinese are committed to pushing US naval and air forces away from the Western Pacific (the South and East China Seas), whereas the US military is determined to stay put. They [the Chinese] see the South China Sea the way American strategists saw the Caribbean in the 19th and early 20th centuries: the principal blue water extension of their continental land mass, control of which enables them to thrust their navy and maritime fleet out into the wider Pacific and the Indian Ocean, as well as soften up Taiwan ... But the Americans will not budge from the Western Pacific.[14]

A recent US Studies Centre report argues that China 'is growing ever more capable of challenging the regional order by force as a result of its large-scale investment in advanced military systems'. It claims that 'Beijing could quickly use limited force to achieve a *fait accompli* victory—particularly around Taiwan, the Japanese archipelago or maritime Southeast Asia—before America can respond, sowing doubt about Washington's security guarantees in the process.'[15] This challenge posed by an assertive and hegemonic

People's Republic of China, that is determined to overturn American strategic primacy in the Indo-Pacific, and revise the existing order into a Chinese-led 'community of common destiny' should be *the* key driver shaping Australia's strategic outlook.[16]

China's naval challenge

If Australia seeks a benchmark challenge against which to plan development of its future defence capability, it must be China's People's Liberation Army (PLA).[17] Australia must match or exceed the PLA in qualitative terms. The 2020 Defence Strategic Update emphasises the need to deter threats at long range and allow credible defensive operations in the event that deterrence would fail. The acquisition of nuclear-powered submarines for the RAN would be justified given this task, the emerging military potential posed by China, as well as the requirement for Australia to accept heavier burdens within the United States-Australia alliance.

The PLA-Navy (PLAN), like the rest of the PLA, is embarking on rapid force modernisation, and closing the qualitative gap between itself and the United States Navy. It is overtaking the United States quantitatively, as well as in long-range stand-off weapons. [18] Key capability developments in the PLA Air Force (PLAAF) and the PLA Rocket Forces (PLARF) as well as PLA Strategic Support Force (PLASSF) in terms of space, cyber and network warfare capabilities add to the growing challenge facing the United States and its allies, including Australia, as China expands its anti-access and area denial (A2AD) capabilities, and begins to build a blue-water power projection naval capability. The PLAN is moving rapidly to becoming a naval force certainly capable of imposing sea control within the first island chain in the 'near seas', and sea denial out to the second island chain in the 'middle seas'.

China is certainly undertaking a very visible shipbuilding effort, but must balance that investment into modern platforms and capabilities with a need to ensure adequate crewing by trained personnel to sustain a much larger and more capable naval force into the future. Goldrick observes that:

> The navy is expanding so fast that personnel planners must be hard-pressed providing crews for the newly completed combatants, let alone filling the inevitable proliferation of staff and technical jobs ashore. The enormity of the challenge has been disguised in part because the

development of the PLA-N is not so much in increased numbers of ships, but in the leaps of capability and sophistication involved. [19]

While it is easy to get fixated on the rapid modernisation of the fleet, China's naval power will also depend on more intangible factors, including recruitment and retention of the right people, and an ability to develop, test and fine-tune effective operational doctrine. Realistic exercises that train the PLAN to fight in a joint and integrated operational environment, and a willingness to practice mission command, are just as important as the numbers and quality of ships and submarines.

China is acutely aware of the importance of these aspects of seapower, and the process of organizational reform of the PLA as a whole that has been underway since 2015 is emphasizing the requirement to 'fight and win local wars under informationised conditions.'[20] That demands not just technological capability, but also an ability to develop operational doctrine to make investment in advanced technology generate decisive military effect.[21] It also demands skilled people to operate advanced military technologies effectively. China is already pursuing development of aircraft carriers and more advanced amphibious vessels, but how are these likely to employed? Are they being acquired with military operations in the 'near seas' in mind, perhaps against Taiwan or in the South China Sea, or does China plan on projecting power and presence into the far seas and oceans, in a manner that would see greater PLAN presence in or approaching Australia's air and maritime approaches?

Considering the future of the PLAN is not just a case of assessing submarines and surface ships—platforms—but more broadly systems for anti-access and area denial (A2AD) operations, and power projection capability. The potential for widespread use of networked autonomous systems, and possibility that new types of technology might render the oceans less opaque, need to be factored into any forward-looking strategic assessment.

China's submarine and undersea warfare forces
The US Navy Office of Naval Intelligence (ONI) has assessed that the number of PLAN submarines—both nuclear and hunter-killer submarines (SSNs and SSKs), as well as ballistic submariens (SSBNs) will grow in coming years, though SSKs will predominate in number over nuclear boats. A recent Report

to the United States Congress on Chinese Naval Modernisation noted that 'China's submarine force continues to grow at a low rate, though with substantially more-capable submarines replacing older units. Current expansion at submarine production yards could allow higher future production numbers'.[22] It noted that ONI predicts that China's submarine force will grow from a total of 66 boats (4 SSBNs, 7 SSNs, and 55 SSs) in 2020 to 76 boats (8 SSBNs, 13 SSNs, and 55 SSs) in 2030. In the same report, the United States Defense Intelligence Agency (DIA), suggests a total of 70 boats by 2030.

A more pessimistic perspective comes from Captain James Fanell. He argues that 'I assess that the PLA-Navy by 2030 will consist of a surface force of over 450 ships, and a submarine force approaching 110 submarines—it may still be a low estimate'. Fanell stresses that 'the most notable feature of our China assessments is that all of our misjudgments have been in the same direction—underestimating China's rise in military aggressiveness and capabilities'.[23] Rick Joe analysed the principal submarine construction yard development at the Bohai Shipbuilding Heavy Industry Company (BSHIC) at Huludao early in 2019 and determined that the yard is being built to support future construction of SSNs and also SSBNs.[24] He argues that the construction yard in fact can sustain construction of up to twelve Type 095 Tang SSNs *simultaneously*, though he notes that it is more likely four SSNs would be built simultaneously. A single SSBN could be built in this facility at any one time. He suggests that 'two SSNs and one SSBN could be launched every year'. If Joe's analysis of two SSNs per year is correct, this assessment would align with ONI's suggestion that 'current expansion at submarine production yards could allow higher future production numbers'. This would imply up to ten Type 095 *Tang* class SSNs could join the PLAN by 2030.

The future PLAN submarine force will therefore see Type 093A and Type 093B *Shang* class SSNs operate alongside the quieter and more capable Type 095 *Tang* class SSN, as well as large number of diesel-electric submarines such as the Type 039 Yuan class SSP. Submarine analyst H.I. Sutton argues that 'at least nine [Type-095 Tang class SSNs] will need to be built to reach the [ONI] 2030 projection', given the need to replace three ageing (and very noisy) Type 091 *Han* class boats. As already noted, the Type 095 SSNs could be built alongside construction of the new Type 096 SSBNs. The Type 096 SSBN will be designed to carry the much longer ranged JL-3 SLBM.[25]

In submarine warfare, silence is golden. China's current submarine capabilities are, however, anything but silent; the future will see much quieter boats. The Type 095 will allegedly be comparable in noise level to early batch *Virginia* class SSNs in the United States Navy, and a vast improvement on the much noisier Type 093A *Shang* class boats.[26] Submarine quieting for the PLAN's SSBNs is even more essential, given their role in providing a survivable second-strike deterrent capability. Even with quieter nuclear boats, it seems unlikely that China would pursue open ocean continuous at-sea nuclear deterrence (CASD) patrols in a manner like Western navies. Adam Ni notes that 'even if the PLAN were operationally able, there are doubts as to whether Beijing is currently ready to make such a major shift in its nuclear posture.'[27] That would imply they would employ a bastion strategy in a manner that is consistent with centralized control of nuclear weapons, geographically concentrated in the Bohai and Yellow Seas, and potentially in the China Sea Basin within the South China Sea.[28]

SSBN Bastions need to be defended, and this key requirement will shape operating patterns—and thus drive capability requirements—for the PLAN in general and, more specifically, for PLAN submarine and undersea warfare capability. Greater numbers of more advanced diesel-electric submarines such as the very quiet Type 039 *Yuan* class SSP (equipped with Air Independent Propulsion), as well as more capable and advanced SSNs would contribute towards defending SSBN bastions, *and* also lend themselves to other vital tasks including supporting PLA joint operations against Taiwan. They could also be deployed south, through key maritime straits, into Australia's air and maritime approaches, and the Indian Ocean, or into the Southwest Pacific. Such deployments would be to support Chinese activity along the Belt and Road Initiative (BRI), to resolve maritime and territorial disputes, sustain Chinese sea lane security and to apply coercive 'sharp power' against ASEAN states, or small states in the Southwest Pacific, in a manner that contributes to Beijing's goal of a compliant community of common destiny.

The submarines themselves will be supported by more advanced undersea warfare (USW) and anti-submarine warfare (ASW) capabilities. These include the successful Type 056A *Jiangdao* ASW corvette, which is now being produced in large numbers.[29] Eli Huang also notes that China is pursuing deployment of fixed acoustic arrays in areas such as the South China Sea.[30]

Add to this Chinese effort in mapping the sea floor and undertaking oceano-graphic research well beyond the South China Sea, including from satellites, to have a better ability to understand oceanic conditions, thus increasing their ability to detect submarine activity. China's ASW capability—long a clear capability gap—is steadily being enhanced.[31]

For Australia, contemplating future submarine operations, both with the existing *Collins* class and future *Attack* class conventional submarines, these developments should be a worrying trend. Unless it is Australia's intention to withhold its submarine forces close to the shore—a step that would under-mine the rationale for acquiring large conventional submarines such as the *Attack* class—the implication must be that the future undersea battlespace is becoming much more challenging if our *Attack* and *Collins* class submarines are to go into harm's way. With the twelfth and final *Attack* class submarine arriving into service in the 2050s, China's PLAN could see potentially 30 advanced SSNs in service based on *current* production capability. By this distant time in the future, China would enjoy successive improvement in submarine quieting, sensor capability and weapons. The performance of future boats will only improve and new types of undersea warfare capabilities could transform the undersea battlespace in unpredictable manners.

As the United States is considering its future with the SSN(X) program, so will China be thinking about successor capabilities to the Type 093B and Type 095 SSNs that might emerge by the mid-2030s. The balance of naval power may be starting to slide in Beijing's favor, in terms of overall number of battle force vessels, and with qualitative gaps narrowing.[32] Geography also acts in China's favour, and its ability to employ large numbers of long-range strike capabilities, including anti-ship ballistic missile (ASBM) systems against vulnerable forward deployed American forces as part of A2AD, is becoming more apparent.[33] In terms of planning Australia's future undersea warfare capability, a key challenge for Australia is that China's growing capabilities will also enable it increasingly to project power closer to Australia.

The Dragon on our doorstep
China is expanding not only its 'A2AD' envelope out to the second island chain, it is developing naval power projection capabilities that reinforce its ability to undertake sea denial within the middle seas and support joint

operations against Taiwan. These same capabilities will also enable power projection along the 21st century maritime silk road into the Indian Ocean and, potentially, along the 'southwest Pacific spur' of China's Belt and Road Initiative (BRI).

This initiative is consistent with Chinese naval thinking originally suggested under former President Hu Jintao's 'new historic missions' in 2004, which has been reinforced by Chinese defence policy under President Xi Jinping in the 2015 Defence White Paper. That document elevated the importance of seapower over the traditional dominance of the land in Chinese military thinking and is of key importance in understanding the current process of Chinese naval modernisation. It states that 'in line with the strategic requirement of offshore waters defense and open seas protection, the PLA Navy (PLAN) will gradually shift its focus from 'offshore waters defense' to the combination of 'offshore waters defense' with 'open seas protection'.

A key paragraph exploring 'force development in critical security domains' reinforces the implications for the changing role of the PLAN.

> The traditional mentality that land outweighs sea must be abandoned, and great importance has to be attached to managing the seas and oceans and protecting maritime rights and interests. It is necessary for China to develop a modern maritime military force structure commensurate with its national security and development interests, safeguard its national sovereignty and maritime rights and interests, protect the security of strategic SLOCs and overseas interests, and participate in international maritime cooperation, so as to provide strategic support for building itself into a maritime power.[34]

Such a stance would be consistent with a requirement by China to protect strategic interests along the 21st century 'maritime silk road' within the Indian Ocean. It is clear that China will not be content to build a naval capability purely for undertaking an offshore counter-intervention strategy within the first or second island chain, as implied by the acquisition of A2AD capabilities.[35] Instead, it will develop naval power projection capabilities to enable it to project naval force well beyond the first and second island chain, along the 21st century maritime silk road within the Indian Ocean. This would be consistent with the acquisition of large, highly capable multi-mission naval surface combatants such as the Type 055 *Renhai* class guided missile cruiser, China's aircraft carrier program that is likely to deliver up to four aircraft carriers to the PLAN, and greater emphasis on amphibious and underway replenishment vessels. It is also consistent with a desire to pursue development of more capable, long-range nuclear powered and propelled submarines such as the Type 093B Shang and, in the future, the Type 095 Tang SSNs.

China's efforts to control the South China Sea through pursuing territorial claims within its self-declared 'nine-dash line' are part of this process.[36] These bases could support PLAN deployments through the South China Sea towards the maritime straits that exit into the Indian ocean and Australia's sea-air gap to the north. The PLA could forward deploy air and naval forces onto its bases on Mischief, Fiery Cross and Subi reefs to allow PLAN surface and submarine forces more easily to project power further away from China, into the Indian Ocean, and Southwest Pacific, or directly against Australia's sea-air gap.

To ensure the ability to project naval power into the Indian Ocean, it would be logical for China to seek to gain control of key maritime straits along the Indonesian archipelago. Analysis in CIMSEC suggests that the key straits are likely to include deep-water straits suitable for submarines to transit submerged, including: the Sulu-Celebes connector between Malaysia and the Philippines; the Makassar Strait; the Lombok Strait; and Ombei-Wetar Strait near East Timor.

In considering Australia's strategic outlook in the next decade it is clear that China will become a blue-water naval power in every sense of the word by 2030, with a far more expansive strategic gaze, and with capability to support its geopolitical ambitions through forward naval forces. The PLAN is clearly looking beyond the 'near seas'—what it refers to as the '*san hai*'—towards deep ocean blue water operations in the Indian and Pacific oceans. [37] The prospect of PLAN forces, including advanced SSNs, operating in Australia's maritime approaches should be a huge cause for concern as the country considers the acquisition of the *Attack* class future submarine. The question must be asked as to whether the *Attack* class submarine—a conventional boat, with lead-acid batteries, will have sufficient operational performance to be 'regionally superior' in the face of this growing challenge?

Technological change in undersea warfare

This analysis looks forward to 2030, but in considering even longer-term possibilities, out to 2050, the challenges become even greater especially when technological transformation of undersea warfare is considered. Although there is always the risk of unexpected strategic shocks and 'black swan' events that could disrupt scenarios based on linear extrapolation from current trends, it seems highly unlikely over the next decade, or beyond, China would willingly reduce the quantity of submarines they deploy or slow the qualitative modernisation of their submarine and surface forces. It is also certain that China will embrace new technology solutions to counter American and allied naval capability.

Exploring the future undersea warfare environment must consider two key technological shifts that could have a transformative impact. The first is the possibility that new sensor, communications, and computer technologies will make the oceans less opaque. The second is clearly a growing role for advanced autonomous underwater vehicles (UUVs). There is also a broader trend towards extending networked command and control into the undersea battlespace.

Unlike the last Cold War, fought when computer and space technology was relatively crude, advances in space and cyberspace now contribute as much to military defeat or victory as does tactical success at sea, control of the air, or dominating the terrain with superior land forces. In the future, satellite systems and the space domain will play a role in submarine detection, alongside more traditional ASW systems, but also contribute towards enhancing the ability of submarines to 'plug and play' with joint networked command and control systems. Space, and the maritime domain will increasingly become interconnected, even below the waves. This opens up new approaches to undersea warfare which raise both opportunity and risk for Australia's naval forces.

In terms of submarine detection, China is already investing in 'LIDAR' (light detection and ranging), and quantum technology systems that are satellite based, and which could make the undersea battlespace less opaque. [38] The report argues that 'an adaptive sensing mesh itself would be built from space platforms' as well as unmanned air, surface and sub-surface systems and emplaced sensor arrays. The use of quantum computing and Artificial

Intelligence (AI), as well as quantum communications, would allow a rapid assessment of an underwater environment. These technologies also open up the possibility that submarines, or large UUVs, will enjoy an ability to 'plug and play' with existing command and control networks more directly.[39] The ability to understand the underwater battlespace with greater fidelity through advanced space-based surveillance capabilities allows greater innovation in thinking about undersea warfare, and greater risk for navies unwilling to embrace these new capabilities quickly. But how might such capabilities be used? Imagine the following scenario.

An Australian *Attack* class submarine has been deployed into the South China Sea to monitor PLA-Navy operations out of Hainan. Its crew are unaware that they have been detected—not by another submarine employing a towed passive sonar array, or by a warship 'pinging' them with an active bow sonar, but from a Chinese satellite orbiting 400 kilometres above the Earth. The satellite is equipped with a 'light detection and ranging' (LIDAR) laser that can detect a submarine several hundred metres below the waves.[40] That information would be communicated at high speed using sophisticated satellite communications to PLA Navy vessels, allowing them to precisely track the RAN submarine and engage it when they so desire. The Chinese may not deploy their own submarine to attack the Australian boat—they may deploy a swarm of extra-large unmanned underwater vehicles (XLUUVs) instead, which are networked via secure quantum communications links with other platforms, or command and control ashore.

With the United States Navy investing in XLUUVs, such as the Boeing Orca platform, and China pursuing similar types of capabilities, the progression of undersea warfare is becoming less predictable and increasingly transformative.[41] The prospect is emerging that crewed submarines—whether nuclear-powered or conventional—will operate in an environment that includes increasing numbers of sophisticated UUVs below the waves, and unmanned surface vehicles (USVs) on the surface. The potential lower cost of acquiring unmanned systems, and the advantages offered by unmanned systems for experimentation and rapid spiral development open up real possibilities for rapid transformation of the future undersea battlespace. The American decision to invest in Orca reinforces their potential to transform undersea warfare capabilities quickly.[42]

It is clear that the Chinese are not going to be left behind American developments. Given the potential naval construction capability already demonstrated for manned submarines, speculating on how quickly China could develop large numbers of sophisticated XLUUVs, which are networked via satellite into quantum command and control systems and, in the future, queued by space or air-based quantum sensors, must raise concerns about the ability of a small number of conventional submarines to survive in contested maritime environments.

Perhaps, a better solution is to invest in highly capable SSNs *as well as* advanced XLUUV platforms to work as a manned-unmanned team, with our swarms of XLUUVs hunting their swarms, protecting our SSNs, whilst hunting theirs. The decision to invest in advanced Integrated Undersea Surveillance Systems (IUSS) and large UUVs in the 2020 Force Structure Plan is a good first step.[43] Australia should join the United States in its Orca XLUUV program to acquire a number of large UUVs to operate alongside crewed submarines. It should also investigate a range of transformational capabilities, including satellite-based systems, quantum technologies and (AI) that could contribute towards enhanced submarine detection. The proposed submarine technology review later this decade will be critical.

If Australia is determined to operate a 'regionally superior submarine' capability, that review needs to debate the case for transitioning from a reduced buy of conventional *Attack* class submarines towards a nuclear powered and propelled submarine capability in the 2040s. For Australia a fleet of advanced nuclear and conventional submarines, complemented by sophisticated XLUUVs and networked undersea warfare surveillance capabilities would more effectively contribute towards responding to a much more challenging and complex undersea warfare environment through to the 2050s.

Endnotes

1   Department of Defence, 2020 Defence Strategic Update, 2020, 1.2—1.3, p. 11.
2   Department of Defence, 2020, 1.7–1.8. p.13.
3   Department of Defence, 2020, 1.9, p.13.
4   The Prime Minister, 'Launch of the 2020 Defence Strategic Update', 1 July 2020, Canberra.

5   Department of Defence, 'Defence strategy responds to changing environment', 2 July 2020, https://news.defence.gov.au/capability/ defence-strategy-responds-changing-environment.

6   Department of Defence, 2016 Integrated Investment Program, 41.5, p. 83.

7   Department of Defence, 2020 Force Structure Plan, 4.7, p. 37; see also Department of Defence, 2016 Defence White Paper, 4.25, 2016; Integrated Investment Program, 4.4., 2016.

8   Michael Shoebridge, 'Australia's new frigates will be fit for the future. Our future submarines should be too', *The Strategist*, 8 October 2019, at https://www.aspistrategist.org.au/ australias-new-frigates-will-be-fit-for-the-future-our-submarines-should-be-too/.

9   Department of Defence, 4.26.

10   HI Sutton, 'Undeclared Submarine Arms Race Takes Hold in Asia', *Forbes*, 18 February 2020, https://www.forbes.com/sites/hisutton/2020/02/18/ undeclared-submarine-arms-race-takes-hold-in-asia/#4d05957145f5.

11   Peter Jennings, 'National Security and Defence Policy', in John Coyne and Peter Jennings (eds), *After COVID-19: Australia and the world rebuild*, vol. 1, ASPI, May 2020, p. 14, https://www.aspi.org.au/report/ after-covid-19-australia-and-world-rebuild-volume-1.

12   Robert D Kaplan, 'How we would fight China', *The Atlantic*, June 2005, https://www.theatlantic.com/magazine/archive/2005/06/ how-we-would-fight-china/303959/.

13   Robert D Kaplan, 'A New Cold War has begun', *Foreign Policy*, 7 January 2019, https://foreignpolicy.com/2019/01/07/a-new-cold-war-has-begun/.

14   Kaplan, 'A New Cold War has begun'.

15   Ashley Townshend, Brendan Thomas-Noone, Matilda Steward, *Averting Crisis: American Strategy, Military Spending and Collective Defence in the Indo-Pacific*, United States Studies Centre, Sydney, August 2019, p. 6.

16   Graham Allison, 'What Xi Jinping Wants', *The Atlantic*, 31 May 2017, https://www. theatlantic.com/international/archive/2017/05/what-china-wants/528561/.

17   The PLA includes not only the PLA Army, but also the PLA Navy (PLAN), PLA Air Force (PLAAF), PLA Rocket Forces (PLARF), as well as PLA Strategic Support Force (PLASSF), PLA Joint Logistics Force and the People's Armed Police.

18   Andrew S. Erickson, 'Chinese Naval Shipbuilding: Full Steam Ahead', *The Maritime Executive*, 18 January 2019, https://www.maritime-executive.com/ editorials/chinese-naval-shipbuilding-full-steam-ahead.

19   James Goldrick, 'China's expanding navy', *The Interpreter*, 11 July 2018, https:// www.lowyinstitute.org/the-interpreter/china-expanding-navy.

20   State Council Information Office, 'China's Military Strategy', Beijing May, 2015, Section III, http://english.www.gov.cn/archive/white_paper/2015/05/27/ content_281475115610833.htm.

21   State Council Information Office, 'China's National Defense in the New Era', 2019, https://www.andrewerickson.com/2019/07/full-text-of-defense-white-paper-chinas-national-defense-in-the-new-era-english-chinese-versions/.

22 Congressional Research Services, 'China Naval Modernisation: Implications for US Navy Capabilities', 21 May 2020, p. 8.

23 James E Fanell, 'China's global Navy eyeing sea control by 2030, superiority by 2049', *Sunday Guardian Live*, 13 June 2020, https://www.sundayguardianlive.com/news/chinas-global-navy-eyeing-sea-control-2030-superiority-2049.

24 Rick Joe, 'Pondering China's Future Nuclear Submarine Production' in *The Diplomat*, 23 January 2019, https://thediplomat.com/2019/01/pondering-chinas-future-nuclear-submarine-production/.

25 Adam Ni, 'The Future of China's new SSBN Force' in Rory Medcalf, Katherine Manstead, Stephen Fruehling, James Goldrick (eds), *The Future of the Undersea Deterrent: A Global Survey*, ANU National Security College, February 2020, p. 29.

26 HI Sutton, 'China's Submarines May be Catching Up with the US Navy', *Forbes*, 24 November 2019, https://www.forbes.com/sites/hisutton/2019/11/24/latest-chinese-submarines-catching-up-with-us-navy/#1a539d1e298c.

27 Adam Ni, 'The Future of China's new SSBN Force', p. 29.

28 China Power, 'Does China have an effective sea-based nuclear deterrent?', CSIS, 19 March 2020, https://chinapower.csis.org/ssbn/; see also Toshi Yoshihara, James R Holmes, 'China's new undersea nuclear deterrent—Strategy, Doctrine, and Capabilities', *Joint Forces Quarterly*, Issue 50, pp. 36–37.

29 Franz-Stefan Gady, 'China's People's Liberation Army Navy Commissions New Type 056A Corvette', *The Diplomat*, 7 January 2020, https://thediplomat.com/2020/01/chinas-peoples-liberation-army-navy-commissions-new-type-056a-corvette/; see also Martin Manaranche, 'China Commissioned its ninth Type 056 Corvette so far in 2020', *Naval News*, 19 June 2020, https://www.navalnews.com/naval-news/2020/06/china-commissioned-its-ninth-type-056-corvette-so-far-in-2020/.

30 Eli Huang, 'China's cable strategy: exploring global undersea dominance', *The Strategist*, 4 December 2017, https://www.aspistrategist.org.au/chinas-cable-strategy-exploring-global-undersea-dominance/.

31 Rick Joe, 'The Chinese Navy's Growing Anti-Submarine Warfare Capabilities', *The Diplomat*, 12 September 2018, https://thediplomat.com/2018/09/the-chinese-surface-fleets-growing-anti-submarine-warfare-capabilities/.

32 Robert Ross, 'The End of US Naval Dominance in Asia', *Lawfare*, 18 November 2018, https://www.lawfareblog.com/end-us-naval-dominance-asia.

33 Michael Evans, 'US would lose Pacific war with China', *The Australian*, 29 June 2020, https://www.theaustralian.com.au/world/the-times/us-would-lose-any-war-with-china-in-pacific/news-story/989d5832d6460e3bd7bbab4ca983967b.

34 State Council Information Office, 2015.

35 Anthony H Cordesman, 'China's New 2019 Defense White Paper', CSIS, 24 July 2019, https://www.csis.org/analysis/chinas-new-2019-defense-white-paper.

36 James Goldrick, 'China's strategic perspective on the South China Sea', *The Strategist*, 26 November 2019, https://www.aspistrategist.org.au/chinas-strategic-perspective-on-the-south-china-sea/; see also James R Holmes, 'Strategic features

of the South China Sea: A Tough Neighbourhood for Hegemons', *Naval War College Review*, vol. 67, no. 2, Spring 2014, https://digital-commons.usnwc.edu/cgi/viewcontent.cgi?article=1317&context=nwc-review.

37   Patrick Cronin, Mira-Rapp-Hooper, Harry Kresja, Alex Sullivan, Rush Doshi, *Beyond the San Hai—The Challenge of China's Blue-Water Navy*, CNAS, Washington DC, May 2017.

38   Roger Bradbury, Scott Bainbridge, Katherine Daniell, et al, *Transparent Oceans? The coming SSBN Counter-Detection Task may be insuperable*, ANU, May 2020, p. 9.

39   John Keller, 'DARPA pushes submarine laser communications technology for ASW operations', *Military and Aerospace Electronics*, 31 January 2010, https://www.militaryaerospace.com/home/article/16723525/darpa-pushes-submarine-laser-communications-technology-for-asw-operations; see also 'Deep secret—secure submarine communications on a quantum level', *Naval Technology*, 5 December 2013, https://www.naval-technology.com/features/featuredeep-secret-secure-submarine-communication-on-a-quantum-level/.

40   'China developing lidar-based satellite to detect deep diving submarines', *Military and Aerospace Electronics*, 1 January 2019, https://www.militaryaerospace.com/home/article/16709692/china-developing-lidarbased-satellite-to-detect-deepdiving-submarines.

41   Bereniuce Baker, 'Orca XLUUV: Boeing's whale of an unmanned sub', 1 July 2019, https://www.naval-technology.com/features/boeing-orca-xluuv-unmanned-submarine/; see also HI Sutton, 'Chinese HSU-001 LDUUV', *Covert Shores*, 2 October 2019, http://www.hisutton.com/Chinese_LDUUV.html.

42   Kyle Mizokami, 'The Navy Might Buy a Bunch of XL Robo-Subs', *Popular Mechanics*, 2 June 2020, https://www.popularmechanics.com/military/navy-ships/a32733280/navy-unmanned-submarine-drones/.

43   Department of Defence, 2020 Force Structure Plan, 4.9, p. 39.

11 Nuclear power for submarines

Denis Mole

*T*he first nuclear powered warship, the submarine USS *Nautilus*, was launched in January 1954. A few hundred nuclear-powered vessels have since been produced, primarily submarines where nuclear power is a natural fit. It allows submarines to remain submerged almost indefinitely. The United States, Russia, China, Britain and France all build and operate nuclear-powered submarines. India has leased a Russian nuclear-powered submarine since the 1980s and is now developing an indigenous nuclear submarine. Brazil and South Korea are developing nuclear-powered submarines. Because nuclear powered submarines are so superior to conventional submarines, the United States, Britain and France do not operate any conventional submarines. The Americans have not built a conventional submarine since the 1950s.

The reasons Australia has not adopted nuclear propulsion are frequently claimed, but never verified. These claims include the refusal of any country to sell Australia nuclear submarines. If this was once true, and I doubt it ever was, circumstances have changed. This claim has not been tested through formal government to government enquiries. Another claim is that Australia does not have a nuclear industry and yet there has been no study to determine what national nuclear industry would be essential, especially in view of modern submarines with reactors that do not require refueling through the life of the vessel. Claims that nuclear submarines are noisy, cannot operate in shallow water or offer only marginal benefits over conventional submarines are myths not supported by professional submariners. My chapter will examine the various competing claims and miscellaneous myths about nuclear submarines and explore current Australian defence policy and potential roles for submarines. I will also review some options that Australia might consider should a future government decide to pursue nuclear power.

Nuclear versus conventional power

Prior to the advent of nuclear power, until the 1950s including both world wars, submarines were designed to operate on the surface most of the time, diving only when in their patrol area to conduct attacks or to avoid nearby anti-submarine forces. During the Second World War, submarines could travel at much faster speeds than most merchant ships that were their primary targets. High speed was achieved by having their diesel engines directly driving their propeller shafts. When submarines did submerge, the diesel engines were disconnected and electric motors propelled the submarines, drawing stored energy from large lead acid batteries. Submarine hull shape was designed for lengthy surfaced voyages and small battery capacity resulted in submarines only being capable of relatively slow speed submerged. A short burst of high speed was possible if the battery was near fully charged. The energy supply then drained rapidly. Consequently, dived conventional submarines are, on the whole, much slower than surface ships.

Development of long range, radar-equipped, anti-submarine aircraft drove conventional submarines off the surface. Towards the end of the Second World War, Germany developed a snorkel mast allowing submarines to be dived, but just beneath the surface, while running diesel engines to recharge their batteries and replenish the air for the crew to breathe. In submariner jargon, this process of recharging batteries via a snorkel mast is known as 'snorting'. Further development of anti-submarine technology progressively rendered snorting submarines vulnerable. First, radar improved so that even exposure of a small periscope above the surface of the sea could be detected. More significant has been the continuing enhancement of passive (listening) sonar capable of detecting noise emanating from a submerged submarine. A dived conventional submarine produces most noise when snorting because it needs to operate diesel engines.

The proportion of time a conventional submarine spends snorting is known as the indiscretion ratio, that is, the submarine is regarded as being very discrete when not snorting and indiscrete when snorting. When a submarine has to transit long distances between its home base and patrol area, a balance has to be achieved between maximum time on station in the patrol area and vulnerability resulting from excessively high indiscretion ratio while transiting to or from that patrol area. As the average speed of the transit increases, so does the indiscretion ratio. A modern conventional submarine transiting long distance at an average speed of 10 knots, typical of a low threat environment, might have an indiscretion ratio of 50 per cent, that is, it would need to spend about 12 of every 24 hours snorting. In a higher threat environment, the average transit speed might be reduced to about six knots with an indiscretion ratio of about 20 per cent. Once in their patrol area, submarines might not need to use much speed and the indiscretion ratio might be less than 10 per cent; if bursts of high speed are used either to pursue targets of interest or to conduct evasion, the indiscretion ratio would increase substantially.

In a nuclear-powered submarine, the uranium fuel fission reaction generates heat which is used to generate steam. The steam drives a turbine generator to produce electric power. Electric motors then power either a propeller shaft or a pump jet. The submarine is therefore freed from any need to expose masts above the surface of the sea to achieve sustained propulsion power. Furthermore, the abundance of electric power enables the submarine to generate oxygen from sea water allowing a constantly suitable atmosphere to be maintained in the submarine for the crew to breathe. Air quality in a nuclear submarine is considerably cleaner than in a diesel submarine and therefore healthier for the crew. Radiation exposure in a modern nuclear submarine is less than the average exposure for the normal civilian population ashore.

Nuclear propulsion allows a submarine to proceed at high speed, higher than almost all surface ships, including surface warships. Conventional submarines have several limiting factors that determine the maximum duration of a submarine patrol, but most significant is the amount of diesel fuel that can be carried. The limiting factor in the duration of a nuclear-powered submarine patrol is crew endurance, typically about three months. Furthermore, because of their much higher, discrete, transit speeds, nuclear submarines spend a much higher proportion of their patrol in their assigned patrol area, not transiting to and from their patrol area. This is particularly germane in Australia's Indo-Pacific situation where likely patrol areas will be at great distances from submarine home base(s). The designated patrol area for nuclear submarines can be changed quite easily while the submarine is on patrol, even if there is great distance from the old to the new patrol area.

Nuclear submarines can be highly effective in open ocean or focal points, whereas conventional submarines are relatively effective only in focal points. This is because if a potential target gets past a conventional submarine just beyond the submarine's weapons range, the submarine does not have the speed and endurance to pursue the target.

Submarines are highly effective anti-submarine platforms, a case of sending a thief to catch a thief. Whilst conventional submarines can be effective against other submarines in focal points, such as in the vicinity of the enemy submarine base, they do not have the speed and endurance to act in direct support escorting transiting surface fleets. Anti-submarine direct support is, however, a major role for nuclear-powered submarines for which conventional submarines are not capable.

While non-nuclear air independent propulsion technologies, such as fuel cells, are available and being introduced in some regional submarines, these are of limited power and endurance, restricting submarine mobility when using this energy source. Non-nuclear, air independent propulsion is generally used whilst loitering in an operating area to reduce the risk of counter detection; it does not improve the submarine's mobility on long transits or overall endurance without refuelling. It does not remove the ultimate reliance on the atmosphere to run diesel generators to charge the battery.

A widespread myth about nuclear-powered submarines is that they are not as quiet as conventional submarines. This is simply not true. When a nuclear-powered submarine is moving at its maximum speed it can be less quiet than a conventional submarine at its maximum speed. It is noisier simply because the nuclear submarine has the power to move at a higher speed than a conventional submarine. When operating at similar speed, nuclear and conventional submarines are equally quiet. It is true that a conventional submarine can get to a quieter state than a nuclear submarine for a limited period by shutting down all ventilation and air conditioning, along with other systems, and cease moving through the water. It can only maintain this level of silence for a few hours. Although a nuclear submarine must keep its reactor functioning, when a conventional submarine is snorting it is considerably noisier than a nuclear submarine operating at similar speed.

Another popular myth about nuclear-powered submarines is that they cannot operate in relatively shallow water. Again, this is simply not true. The minimum depth of water in which a submarine can effectively operate is governed by the dimensions of the submarine, not by the type of propulsion. The current French *Rubis* class nuclear submarine is of similar dimensions to Australia's *Collins* class conventional submarines and the new French *Barracuda* class nuclear submarine will be of similar dimensions to Australia's new *Attack* class conventional submarines. Both the Americans and the British have had entire classes of nuclear submarines that were smaller than Australia's future *Attack* class conventional submarines.

The Australian navy is fortunate to have personnel possessing a good understanding of the benefits of nuclear power. Many former Royal Navy nuclear-trained submariners transferred to the Australian submarine service. Australian submarines frequently exercise with and against United States nuclear-powered submarines and all Australian submarine commanding officers are trained in Europe and are assessed when operating against nuclear submarines. As a result of this experience, it is unlikely any current or recent Australian submarine commanding officer has any doubt about the superiority of nuclear submarines over conventional submarines of similar vintage.

Attack class submarine program

Australia operates six ageing *Collins* class conventional diesel submarines, to be replaced by 12 conventional *Attack* class submarines.[1] The French company, Naval Group, has been awarded a contract to design the *Attack* class and it is intended that Naval Group will also receive a contract for construction of the submarines in Adelaide. Naval Group is building *Barracuda* class nuclear powered submarines for the French navy. Although Australia's *Attack* class submarine has been publicly referred to as the 'Shortfin *Barracuda* class' and will be of similar dimensions to the French *Barracuda* class, they will actually be an entirely new and unique design.

As the *Attack* class submarines are still in the design stage, details of their final characteristics, such as dimensions and delivery schedule are not yet certain. They will not, however, vary significantly from what is already publicly known. The *Attacks* will be approximately 94 metres long, with a submerged displacement of about 5000 tonnes. This will result in their being the largest conventional submarines in the world. The large size is driven by the large quantity of diesel fuel necessary to achieve the range and endurance appropriate to Australia's unique strategic geographic situation. They will have six torpedo tubes and the capacity for either 20 to 30 torpedoes or anti-ship missiles, such as Harpoon. The submarines will be fitted with an upgraded version of the AN/BYG-1 combat system used in American nuclear-powered submarines and Mark 48 MOD 7 heavyweight torpedoes, also as used by the United States Navy. The combat system will be provided by Lockheed Martin.

The *Attack* class submarines will be diesel electric powered with three diesel engines charging large batteries. The submarines will most likely be ordered and built in design batches of three or four submarines per batch. At least the first batch will have conventional lead acid batteries although later batches may incorporate newer battery technology such as lithium-ion. The electric power from batteries will drive an electric motor connected to a pump jet that will propel the submarine through the water.

Construction of the first submarine is expected to commence around 2023–24 and be completed about 2031, followed by a period of first of class trials and operational tests and evaluation. Subsequent submarines will have shorter construction duration, such that they enter service at about two-yearly intervals. The twelfth *Attack* class submarine is not expected to enter service until around 2053. The *Collins* class submarines will be life extended, probably being progressively retired over a decade commencing in the late 2030s. Unless there is a change of direction, Australia will still be building conventional submarines one hundred years after construction commenced of the first nuclear-powered vessel, noting that conventional submarines are inferior, and do not provide the value-for-money capability, that is provided by nuclear-powered submarines.

Australia's Defence Policy—Submarines
The 2009 Defence White Paper stated: 'The Government will double the size of the submarine force (12 more capable boats to replace the current fleet of

six *Collins* class submarines)[2] The quantum of future submarines did not vary from 2009. The 2016 document called for the 12 submarines to be, 'regionally superior submarines with a high degree of interoperability with the United States'[3] This remains government policy. There is no further qualification to which countries this 'regionally superior' capability expectation applies.

Where is Australia's region? The 2016 White Paper frequently refers to the Indo-Pacific region and implies that the entirety of that region applies. For example: 'Our security and prosperity depend on a stable Indo-Pacific region'[4] and 'in the Indo-Pacific region Australia must continue to work with the United States and regional partners to make a positive contribution to security'. The 2016 White Paper further states that: 'In order for Australia and other countries to take advantage of the unprecedented economic growth of the Indo-Pacific region and beyond, we must be willing and able to meet the threats to the peace and stability that has underpinned these positive developments'[5]

In clarifying Australia's need for submarines, the White Paper explains,

> By 2035, around half of the world's submarines will be operating in the Indo-Pacific region where Australia's interests are most engaged. Australia has one of the largest maritime domains in the world and we need the capacity to defend and further our interests from the Pacific to the Indian Oceans and from the areas to our north to the Southern Ocean. Submarines are a powerful instrument for deterring conflict and a potent weapon should conflict occur.[6]

Understandably, China gets frequent special attention in the 2016 White Paper: 'While China will not match the global strategic weight of the United States, the growth of China's national power, including its military modernisation, means China's policies and actions will have a major impact on the stability of the Indo-Pacific to 2035'[7] Again: 'China's Navy is now the largest in Asia. By 2020 China's submarine force is likely to grow to more than 70 submarines'[8]

While couched in appropriate diplomatic nuance, there seems little doubt that China's navy sits firmly within the region for which the Australian Government requires regionally superior submarines.

The changes that have occurred in Australia's region since the 2016 White Paper, particularly regarding China, have been far greater than forecast and of more strategic concern for the Australian Government. This led to the government developing the 2020 Defence Strategic Update. The Strategic Update further clarified Australia's area of strategic interest, noting that 'defence planning will focus on our immediate region: ranging from the north-eastern Indian Ocean, through maritime and mainland South East Asia to Papua New Guinea and the South West Pacific'.[9]

The Strategic Update also noted that 'the prospect of high-intensity military conflict in the Indo-Pacific is less remote than at the time of the 2016 Defence White Paper, including high-intensity military conflict between the United States and China'.[10] It is not that China will necessarily become a threat in the future. China is the country presenting the most serious potential threat at the moment. The Australian government expects China's rapidly expanding force of fast attack nuclear powered submarines to be a factor in the 'regional superiority' required of Australia's future conventional *Attack* class submarines.

The Strategic Update was supported by simultaneous release of the 2020 Force Structure Plan. The Plan did not fundamentally alter the intended development of the submarine force, requiring 12 'regionally superior' submarines. It is nonetheless noteworthy that the security time horizon for the 2016 White Paper is limited to 2035 and the first of Australia's 12 new conventionally powered *Attack* class submarines will not enter service until the early 2030s, with the last boat entering service in the early 2050s and remaining in service until about 2080.

The 2016 Defence White Paper stresses:

> The key strategic requirements for the future submarines include a range and endurance similar to the *Collins* class submarine, sensor performance and stealth characteristics which are superior to the Collins Class, and upgraded versions of the AN/BYG-1 combat system and Mark 48 MOD 7 heavyweight torpedo jointly developed between the United States and Australia as the preferred combat system and main armament.

Notably, there is no stated mention of a need for high speed, particularly sustained high speed which is beyond the capability of conventional power. A conventional submarine can attain high speed but only for a short period of time before it is necessary to recharge batteries or until a fuel source that is independent of the need for air is exhausted.

The limitation on sustained high speed imposes significant restrictions on the area of operations of conventional submarines. The 2016 Defence White Paper explains that 'the key capabilities of the future submarine will include: anti-submarine warfare; anti-surface warfare; intelligence, surveillance and reconnaissance; and support to special operations'. Whereas nuclear-powered submarines are largely effective anywhere, conventional submarines need to operate in focal areas to be effective.

The Second World War demonstrated that submarines are best used offensively, far from home on their enemy's door step. The Germans, Americans and British all ran highly successful submarine campaigns by adopting an offensive strategy, whereas the large submarine forces of Japan and Russia were effectively wasted with too much time spent on a defensive posture. Any expectation that Australia's new *Attack* class submarines will be effective in their roles when operating defensively near Australia is unrealistic. They will be most effective operating in the enemy backyard or at a focal choke point through which the enemy must pass.

Anti-submarine warfare (ASW)

When conducting anti-submarine operations, conventional submarines lack the sustained speed necessary to act in direct support of a surface force. They would be most effective operating in the vicinity of an enemy submarine base where their opposition must by necessity conduct indiscreet activities for sea trials and training. Once an enemy submarine is clear of their home focal area and transiting in open ocean, manned or unmanned maritime patrol aircraft are more effective ASW platforms than conventional submarines. Conversely, nuclear-powered submarines can operate in direct support of an allied surface force to counter enemy nuclear or conventional submarines.

Anti-surface warfare

When conducting anti-surface warfare, conventional submarines lack the sustained speed necessary to pursue enemy surface forces in open ocean. Nuclear submarines are not so constrained, able to pursue or interdict surface forces in both open ocean and focal areas.

Intelligence, Surveillance and Reconnaissance (ISR)

The advantage of using a submarine for ISR is that it can covertly go where other assets cannot. Control of the maritime air and surface environment is not necessary. As long as the submarine remains undetected, the observed targets are likely to continue the activities that most require intelligence collection and/or monitoring. The activities usually occur in an area near the coast of the country that is the ISR target.

In periods of tension a submarine can covertly conduct ISR without escalating the situation, whilst simultaneously being in position to take action if hostilities eventuate. If the presence of the submarine is detected, a conventional submarine is much more vulnerable to counter measures than a nuclear submarine, which has the speed and endurance to evade and clear away rapidly from the area.

Support to Special Operations

The term *Special Operations* refers to delivering or recovering Special Forces, such as Special Air Service Regiment (SASR) troops, behind enemy lines. If Special Forces are confined in a submarine for a long period of time, they can lose fitness. The advantage of nuclear power is that a submarine can transit to and from the target location relatively quickly. Depending on the sea-bed topography, submarines may play a role in amphibious operations, providing landing zone surveillance, ISR and ASW protection for the amphibious force. This form of 'direct support' benefits considerably by the mobility and endurance of nuclear-powered submarines.

Non-nuclear strategic strike—the missing role

There is a significant implicit change in the 2020 Defence Strategic Update. It replaced the Strategic Defence Framework set out in the 2016 Defence White Paper with three new strategic objectives for defence planning. These objectives are to: shape Australia's strategic environment; deter actions against Australia's interests; and, respond with credible military force, when required.[11]

The emphasis on 'deter' and 'respond' should lead to a review of the required submarine capability. Will Australia's future 12 conventional *Attack* class submarines have the capabilities to deter and respond adequately?

Throughout the 20th century, torpedoes were the principal weapons of submarines, except for nuclear-armed ballistic missile submarines (SSBNs). Torpedoes have at all times been the principal weapons of Australian submarines and will be in the new *Attack* class. The Second World War demonstrated that submarines are best used offensively, far from their home base, in areas where there is the greatest concentration of potential targets. The overwhelming majority of submarine targets were not warships, they were merchant ships, either engaged in commerce or logistic support of the war effort. This targeting of merchant ships was not accidental, it was the result of deliberate strategy by the Germans, Americans and British.

All strategic military planners must be cautious against preparing for the last war instead of the next one. The rules of war changed significantly after 1945.[12] International law now requires nations engaged in armed conflict to distinguish combatant targets from civilians. That is not to say that any civilian casualties would be a breach of the law. Harm caused to civilians must serve a concrete and direct military advantage while attacking a legitimate military objective. Damage to civilians and private property must not be excessive.

Changes to the rules of armed conflict alone reduce the likelihood of submarines conducting commerce raiding. Since 1945, the entire structure of commercial shipping has changed, rendering such a role for submarines even less likely. In 1939, 33 per cent of all commercial shipping was British flagged. The majority of the remainder was Japanese or American. Each nation was a participant in the war.[13] Most ships presently sail under flags of convenience, they are often owned by companies headquartered in another country, owned by shareholders from many countries, with the ships and cargoes separately insured by other multinational parties, while not necessarily carrying cargo between any of the countries associated with the ownership or insurers of the ships or cargoes.

In general, therefore, attacking merchant ships will not be a primary role of Australian submarines unless those ships are directly part of a naval or amphibious fleet or are supplying deployed military land and air forces. Besides, whose commerce would Australia attack? Would it be Australian goods being imported into China? Or perhaps exports from China bound for the United States or Europe? If commerce raiding is not viable or realistic in future, this may influence the required torpedo capacity of Australian submarines.

With the advent of nuclear-powered and ballistic missile equipped submarines in the latter part of the 20th century, the primary purpose of nuclear-powered attack submarines (SSN) became anti-submarine warfare, for which torpedoes continued to be the primary weapon, even if not needed in great numbers as was necessary for commerce raiding. Consequently, the number of torpedo tubes in American SSNs reduced from six to four with introduction of the *Los Angles* class in the early 1970s and has continued with the *Seawolf* and *Virginia* classes.

Australian submarines are not in the business of hunting SSBNs. The anti-submarine role of Australian submarines is aimed at countering an opponent's attack submarines which could threaten Australian or allied surface vessels, be they naval or civilian ships. Nevertheless, for an underwater submarine battle, four torpedo tubes would be quite adequate. Six or more torpedo tubes would be preferred if attacking a large surface force or if a mix of weapons such as torpedoes and missiles is required. If, however, the missiles could be launched from other than the torpedo tubes, then four torpedo tubes would be sufficient. This, then, is the crux of the issue: what is or should be the purpose of Australia's submarines?

Ideally, the true value of Australian submarines should be in deterring conflict from occurring in the first place. If deterrence fails then our submarines need a capability that would respond and assist in bringing the conflict to a conclusion, satisfactory to Australia, in the shortest possible time. As noted previously, 'deter' and 'respond' are two of the three 2020 Australian Government strategic objectives.[14] If conflict breaks out with a hostile country that has vast numerical superiority, could Australia's submarines inflict sufficient damage to render the cost too high for the enemy? If Australian submarines are limited to attacking their opponents' naval forces on the high seas, the answer is probably, 'no'. This is especially so if the enemy has a single party totalitarian regime where information flow to their population can be strictly controlled. The cost in terms of enemy lives lost can be kept from the civilian population for a considerable period of time.

To be truly effective as a deterrent and to be able to have a significant impact on the conflict, Australian submarines should be capable of inflicting damage that would be highly visible to the enemy's civilian population, while complying fully with international law. Submarine-launched land attack missiles meet that criteria and should therefore be an essential capability, not just a desirable capability. Legitimate targets would include military installations, fuel supplies, telecommunications, transport infrastructure such as rail and bridges, electric power distribution and some electric power generation. Dams and nuclear power stations cannot be attacked under international conventions. Prohibited targets would include civilian urban residential areas, city business districts of no direct military value, food supplies and storage, hospitals and schools.

Sub-sonic cruise missiles are the current submarine-launched conventional land attack weapon. Future conventional land attack missiles may be ballistic and/or hypersonic cruise missiles. Arming Australia's submarines with land attack missiles would be consistent with extant policy, noting the 2020 Defence Strategic Update stated it is essential that the ADF grow its self-reliant ability to deliver deterrent effects and specifically calls for longer-range strike weapons.[15] While the update calls for long-range land strike weapons, it does so primarily in the section on 'Maritime surface and above water combat' and makes no similar qualification in the section on 'Undersea warfare'.[16] Either the Government has no intention of acquiring land strike weapons for submarines or is being vague, whether intentionally or otherwise.

Launching conventional land-attack missiles from a conventional (diesel) powered submarine is problematic. Too small to include an adequate number of vertical launch tubes, diesel submarines would most likely rely on torpedo tube launch systems. With six torpedo tubes, when approaching a hostile coastline to discharge missiles, submarine commanding officers will most likely keep at least one tube loaded with a torpedo for self-defence, resulting in a salvo of no more than five missiles.

During the past 30 years, the Americans have used submarine-launched cruise missiles in attacks against targets in Iraq, Afghanistan, Syria and Libya. The American tactical practice has been to fire two to four missiles at each target to allow for missile failures, targeting/guidance errors and attrition from the enemy missile defence systems. With only five missiles in a salvo, it is unlikely that a diesel submarine could engage any more than two targets in a single engagement. Any suggestion that tubes could be quickly reloaded to facilitate additional salvoes is unrealistic due to the speed and endurance limitations of diesel submarines. Using space-based detection systems, missile launches from submarines can be detected, pin-pointed and reported in real time.

It is quite likely that by the time some of the *Attack* class submarines enter service, missile detection systems will cover wide areas of ocean approaches to some nations, including the South China Sea and East China Sea, ensuring that the exact location of a submarine firing missiles would be known almost immediately. If Australia's conventional submarines could not effectively operate in these areas, there is almost no likelihood of aircraft or surface ships being able to do so. So where would Australia's deterrence and response come from? The ability to fire a large number of missiles in a relatively short period of time, say less than two minutes, and then clear the firing location at high speed would be desirable characteristics to provide an effective strike and with reasonably high probability of the submarine evading enemy anti-submarine forces. Future Australian submarines should have such missile capability and submarine maneuverability. Such speed/endurance is only achievable with nuclear-powered submarines.

The Americans, British and French each have nuclear-powered and nuclear-armed ballistic missile submarines for their deterrent force, not relying on their attack submarines for that role. Australia is in a different position and must rely on its attack submarines for deterrence. Of the current Western-designed nuclear-powered attack submarines (American *Virginia* class, British *Astute* class and French *Barracuda* class), only the *Virginia* class has large capacity vertical launch tubes for land attack missiles. The latest *Virginia* class submarines (Block V) will have the capability to launch 42 Tomahawk missiles from vertical tubes.

If Australia acquires nuclear submarines at some time in the future, it would most likely occur with the successor designs, noting that HMS *Astute* will be 30 years old in 2040. Therefore, in theory, a future Australian nuclear-powered submarine might be sourced from any of the three Western nations. Although American and British nuclear-powered submarines are currently not for sale, that attitude is changing and would require Australia to package an offer, not just money, to convince them that selling nuclear submarines to Australia would be in their national interest.

Sourcing Australian nuclear powered submarines

Any discussion of Australian nuclear-powered submarines must begin with the power source. The reactors in current nuclear power stations typically generate power in the order of 1000 megawatts. There is now considerable development in progress in several countries regarding Small Modular Reactors (SMR) for electricity generation. SMR are envisioned to generate power from a few megawatts up to 300 megawatts, but on average around 100 megawatts. The small nuclear reactor in submarines currently generate about 100 megawatts and are, therefore, comparable in output to SMRs. The fuel used in the vast majority of nuclear reactors, including all known submarine reactors, is made of uranium oxide (UO_2).

The Americans and the British have adopted a similar approach to submarine reactors. France has adopted a different approach. The United States and Britain use fuel containing a higher percentage of uranium 235, that is, uranium enriched to a higher level, than the fuel used in French submarines. The primary reason the Americans and the British have used the more refined nuclear fuel is that they have evolved their reactor designs so that the fuel core does not need to be replaced throughout the life of the submarine. Their reactors have welded joints and seals which, prior to the current long-life fuel cores, required extremely costly and time consuming nuclear refueling a few times during the life of the submarine.

French submarines, with their less refined uranium fuel, require refueling about once every 10 years. The French submarine reactors have bolted joints and seals, which makes replacing the nuclear fuel core less complex, lower cost and much quicker than the method the Americans and British used prior to full life cores. While there are some benefits to the French model, the American and British approach would appear overall to offer a better 'whole-of-life' approach to nuclear-powered submarine propulsion.

Options for the ADF

The potential options for Australia would be to acquire nuclear submarines designed and possibly built in either France, Britain or the United States. The only remote alternative would be to design and build a unique Australian class of nuclear-powered submarines. Each of these options is examined below.

The French option

The first French nuclear-powered submarines from the early 1970s were designed to carry nuclear weapons, specifically ballistic missiles. It was not until the early 1980s that France built their first nuclear-powered attack submarines. France is currently building the second generation of nuclear-powered attack submarines, the *Barracuda* class. Australia's new *Attack* class conventional submarines have been referred to as the 'Shortfin *Barracuda* class'. This is a misnomer. They will be an entirely distinct class from the *Barracuda* nuclear submarines.

The primary benefits of Australia acquiring French-designed nuclear submarines is, first, that France is the most likely country to agree to sell nuclear submarines to Australia. Second, the company, Naval Group, that builds nuclear submarines for the French navy, is the same company that will design and build Australia's conventional *Attack* class submarines. With Naval Group it should in theory be simpler, in a contractual sense, to abbreviate the *Attack* class program early and switch to building nuclear submarines. Naval Group is also the most likely foreign company to be willing to build nuclear submarines in Australia, with the nuclear reactors manufactured in France. As the first *Barracuda* class submarine was only launched in 2019 and will not be operational before 2021, any French-designed Australian nuclear submarine entering service in the 2040s would most likely be an evolved version of the *Barracuda* class, probably modified to meet specific Australian requirements.

There are disadvantages to a French option, most notably one of language. The Australian navy would ideally send submariners to be nuclear trained in the country of origin of the nuclear power plant. Ideally Australian submariners would also get to 'qualify nuclear' at sea in a submarine of the navy of the country of origin, in this case, France. Language and culture would significantly inhibit such arrangements and experience for Australia.

A further disadvantage might be Australia's requirement to have the same combat system and weapons as the United States Navy. While the Americans have agreed to an American combat system being installed in a submarine being built in Australia by Australian workers employed by a French company, they may be less enthusiastic about their American combat system being installed by French workers in a shipyard in France effectively owned by the French Government. Additionally, Australian submarines built in France would most likely have to go to the United States to conduct post-construction weapons trials, including practice firings.

One further potential difficulty with a French option is that Australia's capability requirements for a nuclear-powered attack submarine might be quite different to that of France, requiring significant design variation from submarines in the French navy. The extent of this variation could not be known until a strategic review is conducted into the future role of Australian submarines. Such a review might conclude that Australia requires submarines to provide a major deterrent effect through the ability to launch a large number of non-nuclear land attack missiles. Such capability would likely involve vertical launch missile tubes, similar to the *Virginia* Payload Module in the USN Block V *Virginia* class submarines.

The fact that French nuclear submarines need to be refueled a few times during the life of the vessel, whereas American and British submarines do not, also needs to be factored into consideration. If Australia does not have a civilian nuclear industry, nuclear fuel for submarine reactors would need to be imported, on multiple occasions for each submarine for French-designed boats. The advantage for France in assisting Australia to acquire nuclear-powered submarines would principally be economic. While maximum economic benefit would accrue if all Australian submarines were built in France, there would still be considerable economic benefit to France for any type of submarine built in Australia by Naval Group due to the French government being the majority shareholder of the company. That is, it would be in France's national interest to build nuclear submarines in Australia rather than risk Australia finding an alternate source.

It might be possible to continue with the *Attack* class conventional submarines, acquiring a smaller quantity, say six boats, while designing and preparing the construction and support facilities for nuclear submarines. If, for example, the first two nuclear submarines were built in France, that could occur in parallel with *Attack* class construction in Australia, with the third and subsequent nuclear submarines built in Australia with no gap in work for Australian industry. Under such an arrangement the first RAN nuclear submarine, built in France, could enter service around 2040 ,and the first Australian-built nuclear submarine in the mid-2040s.

The British option
The United Kingdom has been building nuclear-powered submarines for the past 60 years. The latest British attack type of submarine is the *Astute* class, the first of which was commissioned in 2010. Consequently, any British-designed nuclear submarine for Australia entering service in the 2040s would most likely be a derivative of the successor class to the *Astute* program. The *Astute* class submarines are being built by BAE Systems.

It is not known whether Britain would be willing to build and sell nuclear submarines to Australia and there may be some residual obstacles associated with the conditions imposed by the United States when they provided Britain with its first submarine nuclear reactor in the 1950s. If Britain did agree to sell nuclear submarines to Australia, it might be conditional that they be built in the United Kingdom. A variation of such a condition might be that the first one or two boats must be built in the United KIngdom, with subsequent submarines built in Australia.

It is noteworthy that BAE has a significant presence in Australia and will be building the nine new Australian *Hunter* class frigates in Adelaide. Furthermore, the *Hunter* class building program will be reaching its conclusion at about the time Australia might be seeking its first nuclear submarine.

While Australia would undoubtedly prefer that its nuclear submarines were built in Adelaide, a British build would nevertheless not be entirely negative. If built in Britain, Australian submariners could be trained in the United Kingdom, gain experience aboard Royal Navy submarines and the first one or two Australian boats could operate from the United Kingdom for a period of time while experience is gained and technical support is readily available.

If all submarines were built in Australia, initially submariners could be sent to the United Kingdom for nuclear training. It should be possible to establish an agreement for some Australians to serve in Royal Navy (RN) submarines to gain experience, similar to when the RAN submarine service was established in the 1960s, with a comprehensive exchange program. From an operational perspective, Australian submariners would most easily assimilate with the United Kingdom nuclear training and submarine operating procedures. Many Australian methods for operating procedures were derived from the British when the Australian *Oberon* class submarines were built in the United Kingdom in the 1960s and 1970s. The crewing structure is similar in British and Australian submarines. If the RN and the RAN operated the same class of submarines, a comprehensive personnel exchange program would enable both navies to obtain a broader range of operational experience and should foster improved attraction and retention to both submarine services which is invariably problematic.

If Australia acquired British nuclear submarines and they were all built in the United Kingdom, there would be no great contractual difficulty in transitioning away from Naval Group once the quantity of *Attack* class ordered (less than 12) have been delivered. If, however, continuous build in Australia is required, transitioning from Naval Group Attack class to BAE nuclear submarines would be problematic, but not insurmountable. It would likely be necessary to use much of the *Attack* class construction facility and the Naval Group workforce to transition to BAE, with the two project streams overlapping for a period of about three years.

That British submarines, like American submarines, do not require nuclear refueling during the life of the vessel would be an advantage of the British option ahead of the French option. As with the French option, Australian requirements might require quite significant design variations to British requirements if the earlier argument regarding the need for submarine-launched land attack missiles is accepted. For Britain, there would be considerable economic benefit accruing from any Australian submarines built in the United Kingdom. There would, however, still be economic benefit for the British if Australia acquired the successor to the *Astute* class submarines, even if built in Australia, due to the ability to spread research and development costs across a broader fleet of submarines. Such an arrangement would help to make through life supply chains more economical as well. The British would also benefit from revenue obtained from any nuclear training of Australians in the United Kingdom, as well as other costly training such as the submarine Commanding Officers Qualifying Course. Overall, it would be in Britain's national interest for Australia to acquire the same class of submarine, regardless of where they are built.

The timescale to acquire British-designed nuclear-powered submarines would be similar to the French option, regardless of whether they were built in the United Kingdom or in Australia. The seventh and final *Astute* class submarine is currently under construction. Construction of the successor class, which is more likely relevant for Australia, is likely to commence in the mid-2030s, entering service in the early 2040s. As with the French option, Australia could proceed with the *Attack* class program, albeit with a smaller quantity than currently intended.

The American option

The United States has consistently indicated that it would not sell nuclear powered submarines to any nation, even to a close ally such as Australia. There are some indications that American attitudes have mellowed, at least at the political and civilian bureaucrat level, as well as within the United States Navy. The cost of American submarines might be prohibitive although American submarines would be the best option. In the unlikely event that Australia became involved in armed conflict, it is likely to be alongside the Americans. Operating the same class of submarines would open up a plethora of logistic support options for the platforms and the weapons, options unlikely to be available with French or British classes of submarines.

United States Navy resistance to selling nuclear submarines is generally misunderstood in Australia. It is not that American naval reactors are so advanced and incorporate technology radically superior to the nuclear submarine reactors of other nations. Australia and the United States share much more highly classified intelligence information. Rather than being highly classified because its release to another power would seriously endanger American security, the nuclear reactor technology is highly protected because of fears of the consequences for the United States if another nation failed to operate or maintain an American reactor correctly and a nuclear accident occurred. The American submarine community are particularly worried that, if another country operated submarines with American reactors and an accident happened, it might jeopardise Congressional funding for future United States Navy nuclear submarine programs. Australia would need to package an offer that addresses such concerns.

Using the *Virginia* class purely as an example, what might a 'package offer' look like to the United States? First, the life of all six *Collins* class would need to be extended, resulting in the last *Collins* boat decommissioning around 2047. Producing qualified submariners through the *Collins* class would need to be given national/defence priority. About 100 Australian submariners could be sent to the United States for nuclear training each year. A 'combined' USN/RAN squadron could be established, based in Pearl Harbor at Hawaii, initially with four American submarines while the first four Australian submarines were being built in the United States. As Australian submariners became nuclear qualified, they could initially post into American submarines in the combined squadron so that each American boat has about 75 per cent American and 25 per cent Australian crew. RAN submariners could also filter into the combined squadron staff as well as other United States Navy organisations associated with nuclear safety, refitting, sea training, weapons certifications, missile targeting, and submarine operations management.

As the first four Australian nuclear submarines enter service, they could be crewed with about 75 per cent Australian and 25 per cent American crew and join the combined squadron in Pearl Harbor, which would be under American command although each nation would determine their national submarine deployments on operations. Australia would need to agree that if American submarines with some Australian crew members are required to engage in combat operations, Australia would guarantee not to withdraw its citizens even if Australia were not a party to the conflict. The reverse, however, would not apply. The United States Navy would retain the right to withdraw American personnel from Australian submarines deploying on operations.

The period of time from sending the first Australian submariners for nuclear training until the fourth Australian nuclear submarine became the final unit of the combined squadron would be about 12 years (six years to build the first boat and the three others commissioned at two-yearly intervals). During that 12-year period a range of facilities could be constructed in

Australia. These would include all the facilities necessary at HMAS *Stirling* in Western Australia, a greenfield nuclear submarine maintenance and repair facility, a land attack missile maintenance and storage facility, expanded torpedo maintenance and storage facility and upgrades to weapons/tracking ranges. Australian national organisations would be developed/expanded to oversee nuclear safety with Australia commiting to some continuing nuclear oversight from the Americans in much the same way that the United States Navy conducts periodic audits of the current Australian submarine safety (SUBSAFE) program and certification of the torpedo maintenance facility.

Once the combined squadron is at full strength of eight nuclear submarines (four American and four Australian), the entire squadron could move to HMAS *Stirling*. The benefit to the Americans is that they would have four nuclear-powered attack submarines homeported in the Indian Ocean, saving considerable transit time from current Indian Ocean deployments. Furthermore, they would have ready access to an Indian Ocean submarine repair and maintenance facility as well as weapons replenishment facility, to support any American submarines in the Indian Ocean, not only those homeported in Australia. Once homeported at *Stirling*, command of the combined squadron might rotate USN-RAN every three years. Under a government-to-government memorandum of understanding, these arrangements could be open-ended with an agreed minimum period of say, 15 years, after which Australia could give the United States at least two years' notice if the agreement was to come to an end. The United States could withdraw without notice any time after commencement of homeporting in Australia. The nuclear submarine repair and maintenance facility would be funded and owned by the Australian Government. Managing the facility could be tendered to the American companies that currently support American submarines (who would establish an Australian subsidiary) or through a joint venture with the Australian Submarine Corporation.

At least a further eight Australian submarines would be required to form an Australian squadron. Whether these are built in the United States or Australia would need to be determined and should not influence the decision to go nuclear or not. Sixteen nuclear-powered submarines (four American and 12 Australian) homeported and supported in Australia would generate far more jobs and see more money spent in Australia than would construction and support of a force of 12 diesel submarines.

Under the American option, there might be little point in proceeding with the *Attack* class program, accepting the circa $400 million contract cancellation cost as well as the costs associated with the design contract. There would be little if any design risk with the American option as there would be no need for design changes to meet unique Australian requirements, noting that the *Virginia* class combat system and weapons are already in Australian service.

If the United States agreed to build and sell nuclear-powered submarines to Australia, it might be possible to slot into the build program of later versions (Blocks V, VI and VII) of the *Virginia* class which will continue to be delivered to the United States Navy until the mid-2040s. It is unlikely that, even building in the United States, the first Australian boat could be delivered before the mid-2030s. If the first four submarines were built in the United States, construction of the first Australian built submarine (number 5) would not need to commence before 2035, entering service around 2042, allowing plenty of time to develop the government-to-government agreements, facilities and commercial arrangements.

Hybrid option 1

An alternative, rather than simply accepting a French or British-designed submarine, would be for Australia to be a partner in the design of the submarine, noting it would be the successor class to those submarines currently under construction. This would make it easier to incorporate Australian design requirements, such as a large volume land attack missile capability. The collaborative design option could be Australia-Britain, Australia-France, or all three, with submarine construction occurring in each of the partner nations.

Collaboration between the United Kingdom and France might be difficult to achieve owing to their significant differences in reactor design philosophy. Regardless of whether Australia partnered with the United Kingdom or France, it is unlikely that Australia would acquire its first nuclear submarines through this process until some time in the 2040s. At least six of the conventional *Attack* class would need to be built first.

Hybrid option 2

The likely reasons for American reluctance to sell nuclear submarines to Australia is the risk of not having sufficient control over reactor safety. If this is the principal obstacle, a possible way around it might be to have Australian-American-British collaboration. The result might be the next generation American nuclear-powered attack submarine class, the SSN(X), powered by a Rolls Royce reactor designed and manufactured in the United Kingdom. The submarines could be constructed in Australia, possibly via a joint venture commercial structure. The first United States Navy SSN(X) is not expected to enter service until about 2044. Allowing for operational test and evaluation of the first American boat, it is unlikely that an Australian-built version could or should enter service much before 2050. Consequently most of the planned conventional *Attack* class submarines would be built first, with the nuclear submarines in lieu of the last two or three *Attack* class and then replacing the earlier *Attack* class.

Australian option

The most challenging option would be to acquire uniquely Australian designed and built nuclear-powered submarines, including the nuclear reactors. In the absence of a civil nuclear industry in Australia, it would take many years to build up the nuclear engineering skills and facilities to achieve this option. All 12 conventional *Attack* class submarines would need to be delivered, with development of the civil nuclear industry and the research and development for the first nuclear submarine happening in parallel. If my earlier hypothesis is accepted—that the *Attack* class will not meet the Defence White Paper 'regionally superior' criteria—then the uniquely Australian nuclear option would be problematic. The first nuclear submarine would probably not be available before the mid-2050s and a full force of at least 12 regionally superior submarines would probably not be achieved until about 2080.

Review of future submarine technology

Of the aforementioned options, only the American option might allow a transition directly from the *Collins* class without any conventional *Attack* class submarines, in theory. In reality, it is probably too late to consider nuclear power for submarines replacing the *Collins* class. Nevertheless, the American option should be examined in more detail as an insurance against some unforeseen major obstacle to the *Attack* class program. It is a national strategic imperative that Australia proceeds with the conventional *Attack* class submarines for now. Nevertheless, a submarine capability review should be conducted prior to contracting for construction of all 12 *Attack* class. Such a capability review should thoroughly test the *Attack* class against the 'regionally superior' policy requirement, out to the forecast end of the *Attack* class life cycle, around 2080.

It would likely take at least 15 years and more likely 20-plus years to develop a nuclear industry in Australia. Regrettably, there has been no detailed review as to whether or not a nuclear industry is essential in order to have nuclear-powered submarines. Even if some elements of a nuclear industry are essential or even desirable, if reactors are imported with life-of-type fuel, the level of nuclear industry required in Australia might not be much more than a nuclear waste storage facility. The 2016 Defence White Paper foreshadowed a review of future submarine technology to be conducted in the late 2020s. The Government's Naval Shipbuilding Plan explains that the 12 new *Attack* class submarines are to be the first phase of a Rolling Acquisition Program of submarines, effectively a continuous building program. The future technology review should give serious consideration to nuclear power as a propulsion option for submarines entering service at some time in the 2040s and beyond. As a function of the long lead time to acquire a nuclear submarine capability, the future technology review should be advanced to commence as soon as possible and be concluded no later than 2022.

Transition to nuclear-powered submarines

It would be a great mistake to assume that, as nuclear submarines will not be acquired for at least a few decades, no action is required in the nearer future. The lead time to acquire a nuclear submarine capability will be measured in decades. If there is even a possibility that nuclear submarines may be acquired, policy options need to be explored now.

With the *Attack* class conventional submarine program, Australia might currently be experiencing a strategic mistake of more than a decade ago. It was the 2009 Defence White Paper that first announced government policy to replace the *Collins* class submarines with a force of 12 conventional sub-marines. Three years later, the Defence Minister at the time of that White

Paper, Joel Fitzgibbon, reportedly said that it was a mistake to have ruled out an option for nuclear-powered submarines. If it was a mistake, it was not subsequently corrected, and no serious consideration has been given for nuclear power. The result is that current government policy will have Australia operating conventional submarines until at least 2080.

Due to the age of the *Collins* class, the *Attack* class program should proceed while nuclear options are explored in parallel. Conceding it is almost too late for nuclear-powered submarines to be Plan A, studies should be conducted urgently to see if nuclear power could be a viable Plan B should the *Attack* class program falter for any reason. To do this, dedicated positions need to be established within the Department of Defence, at least within Submarine Capability Branch, the Defence Science and Technology Group and the Capability Acquisition and Sustainment Group.

The possibility of Australia acquiring nuclear-powered submarines should be explored with the American, British and French governments. The reaction of these potential nuclear source countries should be tested at the highest possible level and not simply at navy-to-navy level. If a submarine capability review is followed by a future submarine technology review reporting in 2022, the approach to source governments should ideally occur in parallel with the capability and technology reviews in order to inform the outcome of those reviews.

Should Australia choose to acquire nuclear-powered submarines, even if the full quota of 12 *Attack* class submarines is not built, the time scale to transition to a force of 12 nuclear powered submarines is four or five decades, as described in the various options above. I am not advocating any particular option as all should be explored further. Table 11.1 (below) shows an indicative transition schedule. For the purpose of demonstrating a typical transition plan only, a British model is used. The following assumptions have been made in constructing this indicative schedule:

- all six *Collins* class submarines undergo life extension;
- *Collins* class submarines are withdrawn from service at 24-month intervals from 2038;
- *Attack* class submarines enter service at 24-month intervals from 2032;

- six rather than 12 *Attack* class submarines are built;

- commencing in the late 2020s, Australia and Britain collaborate on the design of a replacement for the *Astute* class nuclear-powered submarines;

- the first successor to the British *Astute* class submarines enters RN service circa 2040;

- from about 2037, some Australian submariners undergo nuclear training in the United Kingdom and gain experience in Royal Navy submarines;

- in the mid-2030s Australia orders two nuclear submarines to be built in the United Kingdom with the first entering service around 2042;

- Australia's first two nuclear submarines each remain based in the United Kingdom for a few years after entering service, qualifying more submariners in nuclear power and gaining operational support experience;

- construction of nuclear submarines in Australia commences about 2040, with the first entering service in 2046; and

- Australian nuclear-powered submarines are built and delivered at 30-month intervals with a 30-year life.

Table 11.1: Example quantity and type of submarines 2031–2070

	2031	2032	2033	2034	2035	2036	2037	2038	2039	2040
Collins	6	6	6	6	6	6	6	5	5	4
Attack	0	1	1	2	2	3	3	4	4	5
Nuclear	0	0	0	0	0	0	0	0	0	0

	2041	2042	2043	2044	2045	2046	2047	2048	2049	2050
Collins	4	3	3	2	2	1	1	0	0	0
Attack	5	6	6	6	6	6	6	6	6	6
Nuclear	0	1 (UK)	1 (UK)	2 (UK)	2 (UK)	3	3	4	4	4

	2051	2052	2053	2054	2055	2056	2057	2058	2059	2060
Collins	0	0	0	0	0	0	0	0	0	0
Attack	6	6	6	6	6	5	5	4	4	3
Nuclear	5	5	6	6	6	7	7	8	8	8

	2061	2062	2063	2064	2065	2066	2067	2068	2069	2070
Collins	0	0	0	0	0	0	0	0	0	0
Attack	3	2	2	1	1	0	0	0	0	0
Nuclear	9	9	10	10	10	11	11	12	12	12

A way forward

Nuclear propulsion for submarines is not new technology. By the time Australia acquires its 12th and final *Attack* class diesel submarine, nuclear propulsion will have been around for a century. When Australia acquired the *Collins* class submarines beginning in the 1980s, it was unlikely that the United States or Britain would have been willing to sell nuclear powered-submarines to Australia and France was only developing its first generation of nuclear submarines.

The 2016 Defence White Paper and, more specifically, the 2020 Defence Strategic Update, leave no doubt that Australia's security outlook contains considerably more risk than when the *Collins* class submarines were acquired. What those documents fail to highlight is that deterioration of strategic circumstances also applies to the United States and that the attitude to assisting Australia with nuclear propulsion may have mellowed.

The nature of submarine warfare has changed. To first deter and, if unsuccessful, respond to hostile aggression against Australia, submarines should be armed with long range land strike weapons. Such weapons need to be carried in large numbers, with the submarine capable of clearing away from the firing location at sustained high speed, capability that only nuclear propulsion can provide.

Australia should proceed to commence the *Attack* class conventional submarine program while simultaneously exploring all options to acquire nuclear propulsion, possibly in lieu of some of the later *Attack* class boats.

The future submarine technology review forecast in the 2016 Defence White Paper should be brought forward to occur not later than 2022. A review of future submarine capability to complement the 2020 Defence Strategic Update and associated 2020 Force Structure Plan should be completed beforehand.

Endnotes

1   2016 Defence White Paper—Exec Sum, p. 19.
2   2009 Defence White Paper—8.39, p. 64.
3   2016 Defence White Paper—Exec Sum, p. 19.
4   2016 Defence White Paper—1.17, p. 33.
5   2016 Defence White Paper—2.4, p. 40.
6   2016 Defence White Paper—4.26, p. 90.

7   2016 Defence White Paper—2.10, p. 42.
8   2016 Defence Strategic Update—2.11, p. 42.
9   2020 Defence Strategic Update—Exec Sum, p. 6.
10 2020 Defence Strategic Update—Exec Sum, p. 14.
11 2020 Defence Strategic Update—2.12, p. 24.
12 Geneva Conventions of 12 August 1949 Additional Protocol of 8 June 1977 relating to the Protection of Victims of International Armed Conflict (Protocol 1).
13 https://www.bbc.co.uk/history/ww2peopleswar/timeline/factfiles/nonflash/a6652091.shtm.
14 2020 Defence Strategic update—2.12, p. 24.
15 2020 Defence Strategic Update—2.22 and 2.23, p. 27.
16 2020 Force Structure Plan—4.6, p. 36.

Part Four

12 Where to from here?

Tom Frame

*T*here are no universally agreed definitions of the national interest or how it is defined. Nor does Australia have a characteristic approach to deciding what serves its interests as a nation. As Australia is a federated commonwealth, the state premiers will present competing if not conflicting views of what it best for their people while the Prime Minister of Australia expects both the Opposition and other non-government parliamentarians to criticise any plan for concerted national action. Perhaps the least elusive area for discerning national interest is defence. The Commonwealth Government has sole constitutional responsibility for the safety the Australian people and the security of their property. Debates about defence usually focus on three things: administration, strategy and procurement. Is the Government managing the portfolio effectively and efficiently? Is it pursuing a strategy able to secure the optimal outcomes in terms of peace and prosperity? Is it acquiring equipment that is best suited to deter would-be aggressors and thwarting the plans of those with hostile intent?

These questions have not changed much since 1901 when the people of the newly formed Commonwealth of Australia wanted an assurance that those they elected were able to discharge the first duty of government—national defence. For the first decade of Australian nationhood, the country relied primarily on Britain until the Royal Australian Navy was established in 1911 and the first of its ships arrived in Sydney Harbour in 1913. Notably, the Australian fleet unit consisted of two submarines. Both *AE 1* and *AE 2* were sunk within the first 12 months of the Great War of 1914–18. Australia acquired a new class of submarine in the 1920s but they did not survive the financial stringencies of the Depression (1929–32). It was not until the

1960s that the country acquired new submarines and made a long-term investment in underwater warfare. Half a century after ordering six *Oberon* class submarines and thirty years after replacing them with the *Collins* class, highly capable submarines remain an integral part of the national defence provision. The most pressing question is what type of submarine propulsion best meets Australia's requirements: diesel electric or nuclear?

The contributors to this collection have argued consistently and, in my view, cogently that Australia ought to consider establishment of a nuclear industry to fulfil domestic power needs and to host a nuclear submarine capability. A shift towards nuclear power would provide electricity that is effectively free from carbon emissions and help to reduce greenhouse gas pollution. Acquisition of nuclear submarines would ensure Australia had the underwater capability needed to ensure an effective response to likely maritime threats.

While a compelling case exists on both fronts, the nuclear option is considered anathema to many policy-makers and, apparently, to many ordinary voters upon whose continuing electoral support a government relies. Despite being early adopters of most new technologies, the effective demonisation of nuclear power in Australia far exceeds community concerns in most other countries, especially advanced G20 economies which have relied for decades on nuclear power in various forms. Mere mention of 'nuclear' appears to unnerve Australians from all parts of the socio-economic-political spectrum. Anxieties about nuclear power are not a product of education or employment.

The word *nuclear* resonates in a manner that engenders fear and fright. There is grave fear of nuclear accidents. They are very rare and have not resulted in mass casualties. Many people are frightened by the proliferation of nuclear weapons although their first and only use was 75 years ago. The word *nuclear* should, therefore, resonate with notions of peace and prosperity. One of the first scientists to recognise that uranium could be harnessed for peaceful purposes was the Australian physicist, Sir Marcus Oliphant. After gaining his undergraduate degree at Adelaide University, he continued his studies at Cambridge and then Birmingham universities where, prior to the Second World War, he conducted research into the peaceful application of energy. In 1941 he was among a small group of scientists to convince the British Government that an atomic bomb was possible. He travelled to the

United States to explain his theory and to confer with like-minded physicists. He subsequently joined the Manhattan Project which developed a weapon that substantially shortened the war in the Pacific. After 1945 and long before being appointed governor of South Australia, Oliphant became an opponent of nuclear weapons—both their development and use. But he remained a strong believer in the peaceful use of nuclear power to generate electricity. He was a humanitarian and a pragmatist.

The explosive content of conventional (that is, non-nuclear) bombs and missiles are manufactured from chemicals but they are not referred to as 'chemical weapons'. The term *chemical weapons* is reserved for specifically prohibited devices designed to kill or maim using poisonous gas, not kinetic explosives. They can be used to cause hurt and harm but chemicals are not inherently insidious. We do not insist on banning the many beneficial applications of chemicals, nor does our society prohibit chemical manufacturing plants in Australia, even though industrial accidents involving chemical plants have led to fatalities and caused illness. Australia is a signatory to international treaties banning chemical weapons and rightly condemns their use anywhere in the world. Plainly, Australians understand the difference between the general use of chemicals and chemical weapons but do not appear to have the same understanding of nuclear power and nuclear weapons.

Part One of this book described some serious deficiencies in public leadership. Australia has produced some inspiring and compelling leaders, yet none has been able to persuade the electorate to give the possible benefits and potential advantages of nuclear power a fair hearing. Prime ministers have, in fact, shied away from leading a debate on nuclear power, fearing an electoral backlash. They have not publicly addressed misinformation and exaggeration despite recognising privately the enormous contribution nuclear power could make to national development. Further, they have perpetuated the inconsistency—one might even say hypocrisy of Australia's attitude to nuclear power—by endorsing export of uranium to other nations and enjoying the protection of friends and allies whose military forces depend upon nuclear power for tactical manoeuvre. The message of Part One is clear: refusing even to consider nuclear power for civilian and military use is senseless if not mindless. If the case for nuclear power is considered and found unconvincing, the nation can pursue other options with conviction and confidence. Given the urgent need to reduce global carbon emission, the nuclear option deserves closer attention than it has received.

In Part Two, some of the more sensational claims about the excessive danger and the high cost of nuclear power are addressed and found wanting. Adopting nuclear power to replace coal for the generation of domestic electric power would enable Australia to ascend the moral high ground on climate change. Investing in nuclear power stations would send an important message to the world, particularly to the small Pacific Island nations most at risk from rising sea levels.

Part Three considered deteriorating security in Australia's foremost area of strategic interest and the coincidental need for nuclear-powered submarines. In providing for national defence, Australia has always relied on quality over quantity. Successive governments have assumed that by operating advanced military technology, Australia can ameliorate the numerical superiority of an opponent in many instances. Equipment used by the Army, the Air Force and the surface elements of the Navy all satisfy that criteria. Not so with submarines. The technology for nuclear submarines has been around for nearly 70 years. Conventional submarines are, on the whole, inferior to nuclear submarines of the same era. Would Australians tolerate RAAF pilots being sent into combat in turbo-prop aircraft when their adversaries have modern jets? Should Australians accept that their grandchildren and great-grandchildren might be obliged to fight a war with inferior equipment, noting that some future crew members of the new *Attack* class submarines will not even be born in the first half of this century? We should not imperil the lives of the nation's young people in submarines that will not be fit-for-purpose for the duration of their service.

What, then, can and should be done? The first priority is achieving a bipartisan, technology-driven, energy policy that appeals to common sentiment and shared aspiration. Australians want secure and stable power supplies and they want the nation's carbon imprint to be reduced. Instead of making energy policy a 'wedge' issue, there is scope for both major parties to pursue a policy that will win electoral endorsement—with the party willing to achieve those ends securing the greatest reward. Plainly, the Labor Party will need to undergo a temperamental philosophical change of heart and at least consider the practical benefits of nuclear power and its potential to create jobs for those more likely to vote Labor. Offering bipartisan support for a creative and imaginative energy policy but excluding the nuclear option is no more than an expression of political risk aversion. Labor must encourage a discussion and be even-handed in debate, allowing science and economics to play a central role.

For its part, the Morrison Coalition Government should embrace the recommendations of the House of Representatives Standing Committee on the Environment and Energy inquiry into the prerequisites for establishing a nuclear industry in Australia. Doing so would not commit the Government to adopting nuclear power. It would simply remove barriers to investigating the possibility thoroughly. I would contend that some of the recommendations do not go far enough. One recommendation is to have ANSTO advise on the technological status of Generation III+ and Generation IV reactors including Small Modular Reactors (SMR). If ANSTO, as the Government's foremost technical advisor, is convinced of the viability of SMRs, the Government should fund a single SMR facility consisting of one or two modules as a technology demonstrator that could be used specifically for research and training. Such a facility could be located near or adjacent to ANSTO and be connected to the national power grid to demonstrate its present capability and future potential for widespread application.

Irrespective of what happens to the Standing Committee's recommendations, there is no barrier to the Government proceeding with nuclear propulsion for submarines. The prohibition contained in *Environment Protection and Biodiversity Conservation Act 1999* relates only to fixed power installations. The Act specifically states that the Minister must not approve an action consisting of, or involving construction or operation of a nuclear fuel fabrication plant, a nuclear power plant, an enrichment plant or a reprocessing facility. While a civil nuclear power industry would naturally complement nuclear propulsion for submarines, it is not essential given that later American and British submarines do not need to be refuelled throughout their operational lives. The one element of a nuclear industry that might be highly desirable is a nuclear waste disposal and storage facility. Such a facility is not, in any case, prohibited by the legislation. As the South Australian Royal Commission on the Nuclear Fuel Cycle (2015–2016) determined, there is a strong and pressing business case for such a facility.

While proceeding with the conventional *Attack* class submarine program, the Government ought to explore nuclear propulsion in a parallel program. The 2020 Defence Strategic Update should lead to immediate review of the projected role of Australian submarines beyond 2040. Such a review would likely conclude that Australia's submarines should provide strategic deterrence through a land strike capability which, in turn, might influence future decisions about the type of propulsion required. The future submarine technology review forecast in the 2016 Defence White Paper could be brought forward and completed by the end of 2022. This timeline would give the Government scope to determine whether 12 conventional *Attack* class submarines are to be acquired or to decide that their transition to nuclear power is required.

High level diplomatic effort might also be applied to explore the attitude of the American, British and French governments to the prospect of selling nuclear submarines to Australia. Prior to making any approach, Australia could develop a package proposal for the United States that included the possibility of some American submarines being homeported in Australia, with an American company operating local repair and maintenance facilities.

These suggestions presume that the Government can and will look beyond the three-year election cycle and consider the very long-term place and importance of submarines in the national defence architecture. Governments sometimes claim that decisions are made firmly in the national interest—which is sometimes code for 'contrary to their party's electoral interests'. Defence is not an area of public policy that usually shapes voting behaviour. Most Australians are usually motivated by material self-interest when casting their votes. They will preference a party whose policies and performance most closely coincide with their personal ambitions and aspirations. Australian governments can make courageous decisions in managing the nation's defence because the consequences do not include losing the next election. And yet both major parties remain highly risk-averse despite having a clear sense of what is at stake. How can Australia compete with nations that are not bound by short political cycles? How will they counter states that take a much more expansive view on all matters?

Nuclear power involves substantial investment in the expectation of securing long-term returns. It is most unlikely that the party opening the door to a more wide-ranging discussion of nuclear power will secure any political benefit from doing so. But the national interest exceeds the lifespan of a government and transcends the longevity of any party. Nuclear power has been around for more than 70 years. The first nuclear-powered submarine was ordered in the early 1950s, followed soon after by power plants to generate electricity.

Australia needs political courage and public leadership to assess the present and potential benefits of nuclear power. There are no insurmountable legislative or technical barriers to Australia deciding its submarines will have nuclear propulsion. Acquiring nuclear-powered submarines that will operate well away from population centres might familiarise the public with the general benefits of nuclear power and lead the way to a comprehensive nuclear industry, an industry that would create jobs and generate revenue. Such objectives are surely in the national interest and worth considering if Australia is committed to preserving its way of life.

✳ ✳ ✳ ✳

www.ingramcontent.com/pod-product-compliance
Lightning Source LLC
Chambersburg PA
CBHW060335100426
42812CB00003B/1002